# 3D PRINTING

# 3D PRINTING

## *An Introduction*

Stephanie Torta

Jonathan Torta

**Mercury Learning and Information**
**Dulles, Virginia**
**Boston, Massachusetts**

Publisher: David Pallai

MERCURY LEARNING AND INFORMATION
22841 Quicksilver Drive
Dulles, VA 20166
info@merclearning.com
www.merclearning.com
(800) 232-0223

This book is printed on acid-free paper in the United States of America.

S. Torta and J. Torta. *3D Printing: An Introduction*
ISBN-13: 978-1683922094

The publisher recognizes and respects all marks used by companies, manufacturers, and developers as a means to distinguish their products. All brand names and product names mentioned in this book are trademarks or service marks of their respective companies. Any omission or misuse (of any kind) of service marks or trademarks, etc. is not an attempt to infringe on the property of others.

Library of Congress Control Number: 2019901183

Our titles are available for adoption, license, or bulk purchase by institutions, corporations, etc. For additional information, please contact the Customer Service Dept. at (800) 232-0223 (toll free). Digital versions of our titles are available at: www.academiccourseware.com and other e-vendors. *Companion files for this title may be requested at info@merclearning.com.*

The sole obligation of MERCURY LEARNING AND INFORMATION to the purchaser is to replace the disc, based on defective materials or faulty workmanship, but not based on the operation or functionality of the product.

# ACKNOWLEDGMENTS

## Stephanie Torta

There were many people who helped with the making of *3D Printing: An Introduction*. Although there are too many to name, I would like to mention a few whose support went above and beyond. First, I would like to say a big, "Thank you," to my brother who is an incredible coauthor. His eagerness to learn, intelligences, and willingness to pass on that knowledge have always inspired me. Writing this book together only increased that feeling. It was a privilege and an honor to work together on this book with him. Thank you to Cara who really should be listed as a coauthor for all her incredible help in bringing this book together. Thank you to Eric, Michelle, Kristin, Kristen, Rachel, Dave, and Colleen for their friendship and insight in the publishing world. To all my friends who gave us their support and let me talk endlessly about the book. To everyone who took the time to share their knowledge with quotes and interviews for the book. And most of all, I would like to give thanks to my mother; without her support this book would not have been possible. Thank you!

## Jonathan Torta

An amazing thank you to all who made this book a reality. I never knew how much went into a book and now I do. To my sister and her patience who was the main mover of this entire enterprise going far beyond author and touching every aspect of this book and had to deal with me learning on the fly. To my wife Cara and mother Diana who corrected the many typographical mistakes and wrangled the words far beyond my own capability and much more. To my good friends Jen, Dave, and Dale that not only gave me support but also interesting projects and stories to add to this book. They keep me on my toes and push my limits with their interesting ideas and projects. And to Eric and other fellow 3D printers who were kind enough to give us insights on how they use these amazing devices in their businesses and also shared their stories. This book was a much larger project then I ever thought it would be and it required all to complete it. Thank you!

# CONTENTS

# INTRODUCTION

## Introduction from the Maker

I have been 3D printing for just under 10 years, starting from a kit and moving to quite a few other printers along the way, always learning more and modifying them along the way. Being a tinkerer, hobbyist and DIYer, 3D printing was a perfect match, allowing me to make useful parts and also being a hobby all in its own right. I am always learning more, but I've found that others have been coming to me for help and advice, and that my experience was valuable to the community. So, I helped out where I could on various forums and enthusiast groups. In making this book, I'm hoping to help many more people by gathering often repeated tips in a single place. In this process I also learned quite a bit in the 3D printing, authoring and bookmaking realms. I very much hope everyone that reads this book can find something useful and helpful.

## What this book is trying to achieve

*3D Printing: An Introduction* is intended to give aspiring makers a realistic look at the overall 3D printing process. Many books have been written about select aspects of 3D printing, this book tries gives the maker a true understanding of what it is actually like to work on a project, whether it be by using a 3D printing service or using their own printer. And to help new makers navigate potential pitfalls when printing.

## What this book entails

The first part of *3D Printing: An Introduction* focuses on an overview of the 3D printing process and how it fits into our lives. What types of printers, services, materials, and software is available

The second part of this book is more focused learning. It shows the first steps and what to think about when setting-up and calibrating your printer for your first print. We also gave some practice challenges and information on finishing and refinement.

The third part of the book expands on the basic printer as well as resources to increase your knowledgebase.

**IN EXTRAS**

In addition, we have provided a DVD and a site for online download that includes a number of additional videos, images, a quick reference print checklist, a useful link guide, and practice files. We also included a few additional learning challenges and exercises for further study. The print checklist is useful to post near your printer for a quick reference guide. The companion files on the disc may be downloaded by writing to the publisher at info@merclearning.com.

# IMAGE LINKS AND LICENSING

All images are copyright by Jonathan or Stephanie Torta unless otherwise noted.

Part 01:      By ssp48 - 3d-printer-for-plastic-L4K3BJW - Envato Elements: https://elements.envato.com/

Chapter 01:  By ssp48 - 3d-printer-for-plastic-PNLVJNN - Envato Elements: https://elements.envato.com/

Figure 1.1:   By Svitlana Lozova [CC BY-SA 4.0 (https://creativecommons.org/licenses/by-sa/4.0)], from Wikimedia Commons: https://commons.wikimedia.org/wiki/File:3D_printing_functional_prototypes.jpg

Figure 1.2:   By Jonathan Juursema [CC BY-SA 3.0 (https://creativecommons.org/licenses/by-sa/3.0)], from Wikimedia Commons: https://commons.wikimedia.org/wiki/File:Felix_3D_Printer_-_Printing_Head_Cropped.JPG

Figure 1.3:   By Creative Tools from Halmdstad, Sweden [CC BY 2.0 (https://creativecommons.org/licenses/by/2.0)], via Wikimedia Commons: https://commons.wikimedia.org/wiki/File:3D-print_of_a_spool_holder_on_a_Printrbot_Simple_Metal_3D-printer_(15646389161).jpg

Figure 1.4:   By Maurizio Pesce from Milan, Italia (Ultimaker 3D Printer) [CC BY 2.0 (https://creativecommons.org/licenses/by/2.0)], via Wikimedia Commons: https://commons.wikimedia.org/wiki/File:Ultimaker_3D_Printer_(16862205332).jpg

Figure 1.6:   By Kaboldy [CC BY-SA 3.0 (https://creativecommons.org/licenses/by-sa/3.0)], from Wikimedia Commons: https://commons.wikimedia.org/wiki/File:STL_sample_2.png

Figure 1.8:   By PranjalSingh IITDelhi [CC BY-SA 4.0 (https://creativecommons.org/licenses/by-sa/4.0)], from Wikimedia Commons: https://commons.wikimedia.org/wiki/File:Supports_in_3D_printing.png

Figure 1.9:   By Jonathan Juursema [CC BY-SA 3.0 (https://creativecommons.org/licenses/by-sa/3.0)], from Wikimedia Commons: https://commons.wikimedia.org/wiki/File:Felix_3D_Printer_-_Printing_Set-up_With_Examples.JPG

Figure 1.10:  By John Abella from Outside Philadelphia, USA (Make Magazine 3D Printer Shootout) [CC BY 2.0 (https://creativecommons.org/licenses/by/2.0)], via Wikimedia Commons: https://commons.wikimedia.org/wiki/File:3D_print_in_process_(9437659715).jpg

Figure 1.11:  Mos.ru [CC BY 4.0 (https://creativecommons.org/licenses/by/4.0)], via Wikimedia Commons: https://commons.wikimedia.org/wiki/File:3D_printed_tactile_replica_of_the_Tsar_Cannon.jpg

Figure 1.13:  By monkeybusiness - female-college-student-printing-3d-object-in-PJDXM2Z - Envato Elements: https://elements.envato.com/

Chapter 02:  By ssp48 - 3d-printer-for-plastic-4QR9WVP - Envato Elements: https://elements.envato.com/

Figure 2.1:  By Intel Free Press [CC BY-SA 2.0 (https://creativecommons.org/licenses/by-sa/2.0)], via Wikimedia Commons: https://commons.wikimedia.org/wiki/File:Mojo_3D_Printer.png

Figure 2.1:  NASA/MSFC/David Olive. - https://www.nasa.gov/centers/marshall/news/nasa-advances-additive-manufacturing-for-rocket-propulsion.html

Figure 2.3:  NASA/Glenn Benson. - https://images.nasa.gov/details-KSC-20180316-PH_GEB01_0132.html

Figure 2.4 – By Creative Tools from Halmdstad, Sweden [CC BY 2.0 (https://creativecommons.org/licenses/by/2.0)], via Wikimedia Commons: https://commons.wikimedia.org/wiki/File:VIUscan_handheld_3D_scanner_in_use.jpg

Figure 2.5 – By Ryan Somma from Occoquan, USA (Fossils and Foam) [CC BY 2.0 (https://creativecommons.org/licenses/by/2.0)], via Wikimedia Commons: https://commons.wikimedia.org/wiki/File:3D_printed_Spinosaurus_skulls.jpg

Figure 2.6 – By Svitlana Lozova [CC BY-SA 4.0 (https://creativecommons.org/licenses/by-sa/4.0)], from Wikimedia Commons: https://commons.wikimedia.org/wiki/File:Architectural_model_printed_with_an_Ultimaker_3D_printer.jpg

Figure 2.7 – By Marczoutendijk [CC BY-SA 4.0 (https://creativecommons.org/licenses/by-sa/4.0)], from Wikimedia Commons: https://commons.wikimedia.org/wiki/File:3D_printed_concrete_bicycle_bridge_in_Gemert_(NL).jpg

Figure 2.8 – By 3DPrinthuset (Denmark) [CC BY-SA 4.0 (https://creativecommons.org/licenses/by-sa/4.0)], via Wikimedia Commons: https://commons.wikimedia.org/wiki/File:The_3D_construction_printer_constructing_The_BOD.jpg

Figure 2.9:  By 3DPrinthuset (Denmark) [CC BY-SA 4.0 (https://creativecommons.org/licenses/by-sa/4.0)], via Wikimedia Commons: https://commons.wikimedia.org/wiki/File:The_building_on_demand_(BOD)_printer.jpg

Figure 2.10:  By 3DPrinthuset (Denmark) [CC BY-SA 4.0 (https://creativecommons.org/licenses/by-sa/4.0)], via Wikimedia Commons: https://commons.wikimedia.org/wiki/File:The_BOD_-_3D_printed_walls_of_the_structure.jpg

Figure 2.11:   By 3DPrinthuset (Denmark) [CC BY-SA 4.0 (https://creativecommons.org/licenses/by-sa/4.0)], via Wikimedia Commons: https://commons.wikimedia.org/wiki/File:The_BOD_-_Europe%27s_first_3D_printed_building.jpg

Figure 2.12-15:   Use with permission from Natural Machines: https://www.naturalmachines.com/

Figure 2.16:   A. Envato Elements: broccoli-PPDA48C - https://elements.envato.com/ - sabinoparente  B. Use with permission from Natural Machines: https://www.naturalmachines.com/

Figure 2.17:   By Maurizio Pesce from Milan, Italia (ChefJet Candy 3D Printer) [CC BY 2.0 (https://creativecommons.org/licenses/by/2.0)], via Wikimedia Commons: https://commons.wikimedia.org/wiki/File:ChefJet_Candy_3D_Printer_(16675977970).jpg

Figure 2.18:   By grafvision - dental-prosthesis-PYR6RKL - Envato Elements:  https://elements.envato.com/

Figure 2.19:   By Grey_Coast_Media - doctors-hands-holding-silicone-mouth-guard-PGUYKJ5 - Envato Elements: https://elements.envato.com/

Figure 2.20:   By Maikel Beerens, Xilloc [CC BY-SA 4.0 (https://creativecommons.org/licenses/by-sa/4.0)], from Wikimedia Commons: https://commons.wikimedia.org/wiki/File:Xilloc_Patient_Specific_Implant_Titanium.png

Figure 2.21:   By The U.S. Food and Drug Administration (3-D Printed Prosthetic Hand - blue (5229)) [Public domain], via Wikimedia Commons: https://commons.wikimedia.org/wiki/File:3-D_Printed_Prosthetic_Hand_-_blue_(5229)_(18492491235).jpg

Figure 2.22:   By CSIRO [CC BY 3.0 (https://creativecommons.org/licenses/by/3.0)], via Wikimedia Commons: https://commons.wikimedia.org/wiki/File:CSIRO_ScienceImage_1761_3D_printed_titanium_horseshoes.jpg

Chapter 03:   By Pressmaster - new-device-P9ULWLQ - Envato Elements: https://elements.envato.com/

Figure 3.1:   By dolgachov - happy-children-with-3d-printer-at-robotics-school-P248X6F - Envato Elements: https://elements.envato.com/

Figure 3.2:   Used with permission from Makers Empire

Figure 3.3:   By Libraries Taskforce (Fab Lab in Exeter Library) [CC BY 2.0 (https://creativecommons.org/licenses/by/2.0)], via Wikimedia Commons, https://commons.wikimedia.org/wiki/File:Fab_Lab_in_Exeter_Library_(19744255694).jpg

Figure 3.4:   By Texas State Library and Archives Commission from Austin, TX, United States (Testing out the 3D printer) [CC BY 2.0 (https://creativecommons.org/licenses/by/2.0)], via Wikimedia Commons: https://commons.wikimedia.org/wiki/File:Testing_out_the_3D_printer_(42000207351).jpg

Figure 3.5:  Used with permission from Dan DeVona

Figure 3.6:  Used with permission from Makerbot

Figure 3.7:  By Libraries Taskforce (Fab Lab at The Word) [CC BY 2.0 (https://creativecommons.org/licenses/by/2.0)], via Wikimedia Commons, https://commons.wikimedia.org/wiki/File:Fab_Lab_at_The_Word_(33923572260).jpg

Chapter 04:  By stokkete - 3d-printing-and-education-P6UENC3 - Envato Elements: https://elements.envato.com/

Figure 4.8–9:  Used with permission from Eric Johnson

Figure 4.13–17:  Photo image courtesy of Jamber, Inc.

Figure 4.18–21:  Photo image courtesy of Lucas Phillips

Chapter 05:  By Pressmaster - using-innovation-PKWV69D - Envato Elements: https://elements.envato.com/

Figure 5.2:  By Math buff [CC BY-SA 4.0 (https://creativecommons.org/licenses/by-sa/4.0)], from Wikimedia Commons; https://commons.wikimedia.org/wiki/File:Cartesian_xyz_ijk_coordinates.svg

Figure 5.4:  By Z22 [CC BY-SA 4.0 (https://creativecommons.org/licenses/by-sa/4.0)], from Wikimedia Commons; https://commons.wikimedia.org/wiki/File:Large_delta-style_3D_printer.jpg

Figure 5.5:  By Svjo [CC BY-SA 3.0 (https://creativecommons.org/licenses/by-sa/3.0)], from Wikimedia Commons: https://commons.wikimedia.org/wiki/File:Polar_coordinate_system-2.png

Figure 6.2:  By Maurizio Pesce from Milan, Italia (3D Printing Materials) [CC BY 2.0 (https://creativecommons.org/licenses/by/2.0)], via Wikimedia Commons: https://commons.wikimedia.org/wiki/File:3D_Printing_Materials_(16837486456).jpg

Chapter 07:  By Dirk van der Made [GFDL (http://www.gnu.org/copyleft/fdl.html) or CC BY-SA 4.0 (https://creativecommons.org/licenses/by-sa/4.0)], from Wikimedia Commons: https://commons.wikimedia.org/wiki/File:Schuifmaat_bottom_mechaniek.png

Figure 7.2:  By Polygon data is BodyParts3D. (BodyParts3D) [CC BY-SA 2.1 jp (https://creativecommons.org/licenses/by-sa/2.1/jp/deed.en)], via Wikimedia Commons, https://commons.wikimedia.org/wiki/File:Cervical_vertebrae_from_BodyParts3D_on_MeshLab.png

Figure 7.3:  By Svitlana Lozova [CC BY-SA 4.0 (https://creativecommons.org/licenses/by-sa/4.0)], from Wikimedia Commons: https://commons.wikimedia.org/wiki/File:Cura_software.jpg

Figure 7.14:  Photo image courtesy of Jamber, Inc.

# PART 1

# 3D Printer Overview

# CHAPTER 1

# A Brief Overview of 3D Printing

## OVERVIEW AND LEARNING OBJECTIVES

**In this chapter:**

- 1.1 – What is 3D Printing?
- 1.2 – What are the steps of the printing process?
- 1.3 – What are some common parts used in a desktop 3D printer?
- 1.4 – How was the technology for 3D Printers invented and first used?
- 1.5 – What is the potential for 3D printers?

## › 1.1 – What is 3D Printing?

When the word *printer* is used, most people think of a conventional printer they might use at home or in the office to print out text and images on paper. These printers print in a flat two-dimensional (2D) space using the dimensions length and width. A *three-dimensional (3D) printer* uses length and width but also adds depth to the print (see **Figure 1.1**). This transforms a flat print into a tangible, workable object you can hold and use.

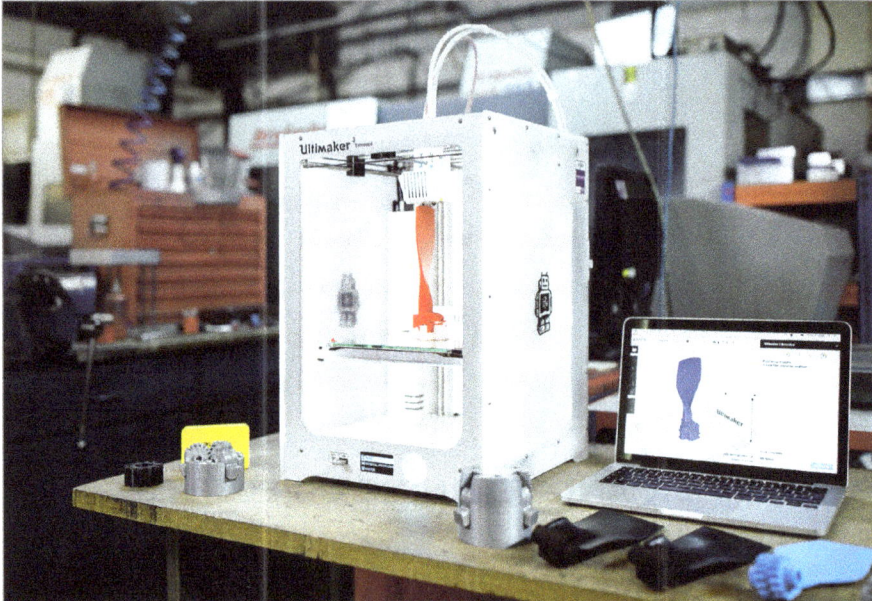

FIGURE 1.1 – A number of 3D printed functional prototypes sliced in Ultimaker Cura and 3D printed on Ultimaker 3 Extended with PLA and water-soluble PVA. Svitlana Lozova.

---

**BRIEF REVIEW OF TERMS**

**Three-dimensional (3D) Printer** is a manufacturing tool that creates physical objects from a 3D model design using an additive manufacturing method that adds layers upon layers of material.

---

Envision the process as if you are printing a flat circle on a piece of paper lying on a table, and then you pull that printed circle "up" from the flat surface creating a physical cylinder. See **Figure 1.2**, which illustrates how a spire rises up from the flat surface during the 3D printing process.

FIGURE 1.2 – The printing head of a FELIX 3D Printer in action. Jonathan Juursema.

3D prints can take almost any form, depending on the size of the printer. After the initial printing process is completed, 3D prints can be linked or fused together to form larger objects. **Figure 1.3** shows a complex object with individually printed parts in a variety of shapes that were assembled after printing.

*Additive Manufacturing* is a general term referring to a variety of fabrication processes that use a manufacturing tool to create a physical 3D object by add-

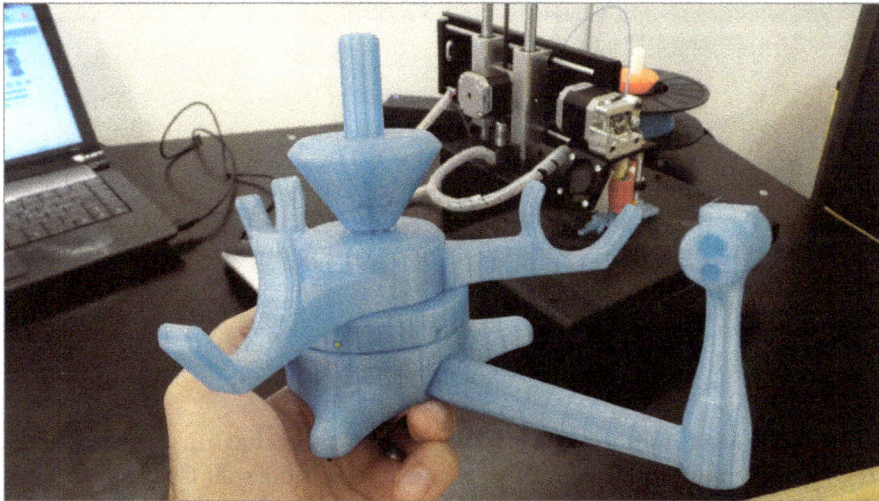

FIGURE 1.3 – 3D print of a spool holder on a Printrbot Simple Metal 3D-printer. Creative Tools.

ing material. A 3D printer is one subset of this type of manufacturing process because it continuously adds layer upon layer of material to build a physical 3D object.[1] This is different from *subtractive manufacturing* that takes material away from existing resources to create an object, or the *consolidation processes* that take smaller parts, combines them together, and fuses them to create the designed object.

At its core, 3D printing is a manufacturing method that takes a digital design and creates a physical 3D object by building up layers of a selected material. In **Figure 1.4**, we see a printer adding material to create a robot layer-by-layer, with some finished printed robots to the right.

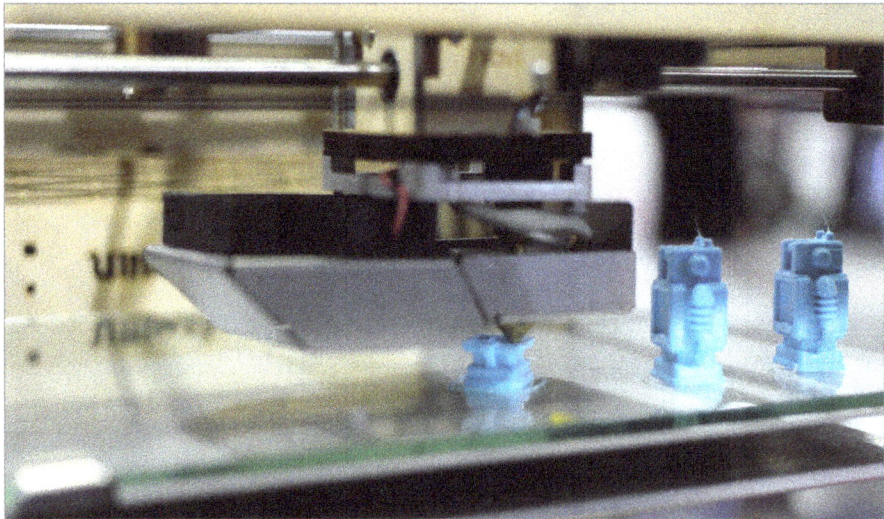

FIGURE 1.4 – Ultimaker 3D Printer. Maurizio Pesce.

## BRIEF REVIEW OF TERMS

**Additive Manufacturing** is an overall term referring to a variety of fabrication processes that uses a manufacturing tool to create a physical object by adding layer upon layer of material. 3D printers are one subset of this type of manufacturing process. (like paper mache)

**Consolidation Processes** refers to a manufacturing method that uses the fusion of smaller individual parts or objects to bond and create a new object from a design. (like a brick wall)

**Subtractive Manufacturing** is a fabrication process that cuts or drills away from a solid material block to create an object. Industrial CNC (Computer Numerical Control) machining is a subset of this process. (like carving)

3D printers were first used as an innovative tool for industrial needs, most notably *rapid prototyping*. One of the greatest advantages of using a 3D printer is its ability to modify or create a unique object that you cannot obtain easily or that does not exist. This makes the use of 3D printing ideal for certain types of prototyping for tools and machine parts that in the past required molds.

Early printers were large and expensive to operate. With the advancements in technology, the costs of a 3D printer have decreased while the available sizes and functionality have increased. The types of materials and *filaments* have also seen advances and have greatly increased the variety of items created using a 3D printer. This has allowed greater use of 3D printers at the small business and consumer market levels.

### BRIEF REVIEW OF TERMS

**Filament** is the raw thermoplastic material used in certain types of 3D printers.

**Rapid Prototyping** is a method that turns a digital design into a physical object using a quicker process, such as 3D printing, rather than traditional methods.

Today, desktop 3D printers can be found in homes, schools, public libraries, retail stores, and businesses along with an increased use in industrial manufacturing.

See Chapters 2 through 4 for how 3D printers are being used. See Chapter 6 for information on the types of materials and filaments available.

## › 1.2 – What are the steps of the printing process?

### Process overview

There are many ways to start a 3D print and we will go more in-depth on the 3D printing process in Part 2 of this book. However, in this section we will briefly describe the process it takes to create a 3D print. These steps generally remain the same for most types of 3D printers and are the cornerstones of 3D printing. **Figure 1.5** shows a diagram depicting the steps and how they are related.

Obtaining a 3D *model* and refining the print are steps that have multiple methods of completion based on your project or equipment. The original source matter

```
        3D scanners        Online download
             ↓                    ↓
Computer programs ●→ **Obtaining a 3D Model** ←● Parametric methods
                           ↓
                 **Selecting a File Format**
                           ↓
                 **Slicing the 3D Model**
                           ↓
                        **Printing**
                           ↓
     Painting ●→ **Refining the Print** ←● Assembly
                           ↑
                 Smoothing the layers
```

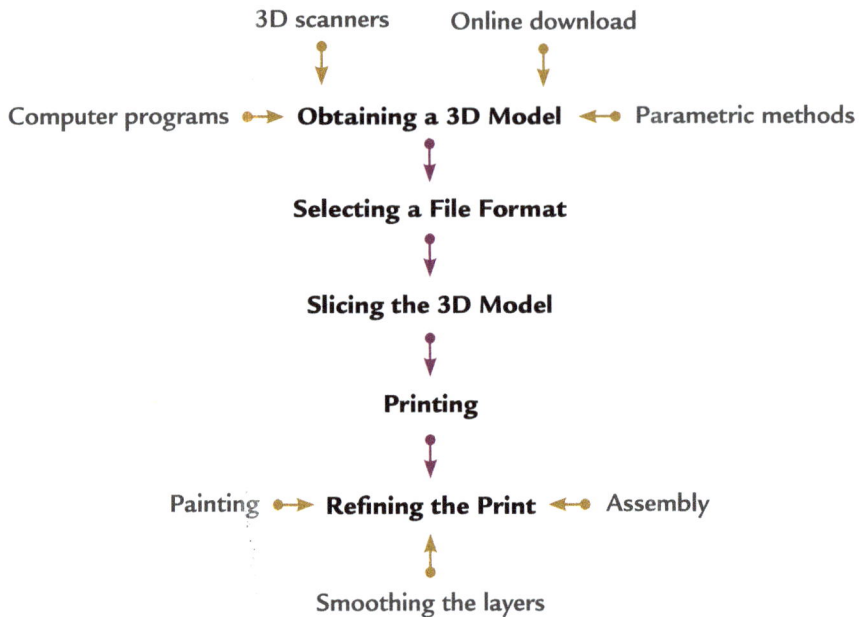

FIGURE 1.5 – The 3D printing process.

determines how a 3D model is created or obtained, while the amount of cleaning and refinement determines the methods of finishing.

**BRIEF REVIEW OF TERMS**

**Model** or Modeling is a digital design created by CAD software, 3D scanner, or digital camera.

MAKER'S
NOTE

The method of creating the 3D model and finishing your print will vary depending on your project.

## Obtaining a 3D Model

The 3D printing process starts with a digital design, or model. This model is a digital 3D representation of a solid object consisting of triangles. The surfaces of these triangles are stored in the computer file to describe the geometry of the model. All 3D objects and formats used in 3D printing use triangles to define the surface of a 3D model (see **Figure 1.6**).

There are many ways to create or obtain a 3D model like using computer programs, scanners and camera, or parametric mathematical equations.

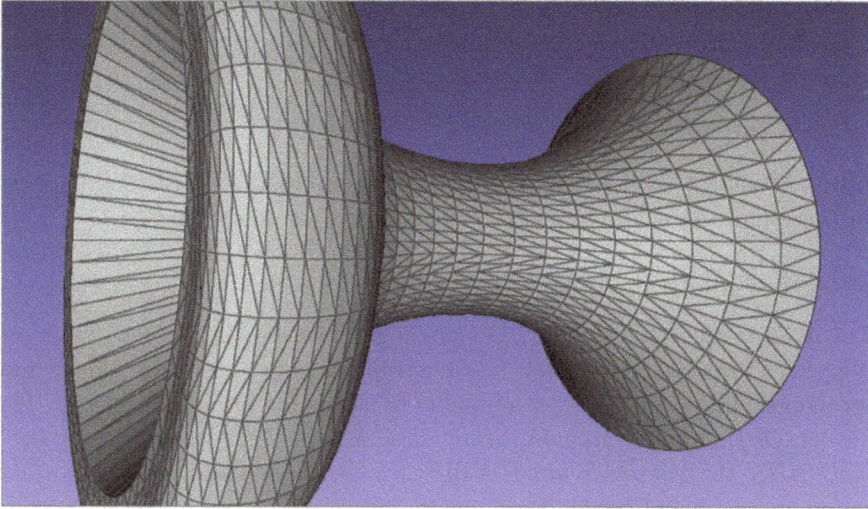

FIGURE 1.6 – STL sample. Kaboldy.

- 3D models can be created by designing the object within 3D animation, art, gaming, or *computer aided design (CAD)* software.
- 3D scanners can scan the entire surface and sides of an object to create a 3D model. This includes digital cameras that take a 2D image and reconstruct and render to 3D using selected computer software.
- *Parametric* methods using mathematical equations can also be used to create 3D models. This method is sometimes used by online sites.
- Premade 3D models can also be downloaded from sites on the internet for both commercial and private use.

An example of obtaining a 3D model is when a *maker* wants to create a lens cap holder for their camera: they have a few options to obtain or create a 3D model. The maker could take measurements of the lens cap and camera strap and design the object using 3D modeling software. The maker could also search online for a premade model at the correct size or one that is modifiable.

## BRIEF REVIEW OF TERMS

**Computer Aided Design (CAD)** is software that assists in creating or modifying a digital model or design.

**Maker** is the person who creates the 3D print.

**Parametric** refers to a mathematical equation that represents the points of an object.

**Web link**

This web link is an example of a premade lens cap holder a maker can download and use. Additionally, this example has a parametric object that can be customized.

https://www.thingiverse.com/thing:2004739

After obtaining a 3D model from one of these sources, the information must be converted into a file format that your slicing program understands.

## Selecting a File Format

The most common file type for 3D modeling information is the *Stereolithography (STL)* format. You will likely encounter this file type when searching online for models. However, there are a number of usable file formats like: *Object file format (OBJ)*, *Additive Manufacturing File (AMF)*, and *3D Manufacturing Format (3MF)*. We will review some of the advantages and disadvantages of using different formats in Chapter 8.

### BRIEF REVIEW OF TERMS

**3D Manufacturing Format (3MF)** is an XML-based data format developed and published by the 3MF Consortium. It is used to store model information like shape, color, and material.

**Additive Manufacturing File (AMF)** is an XML-based data format that stores model information like shape, color, and material.

**Object file format (OBJ)** is an older 3D model format used with various 3D editors.

**Slice** or Slicing is the act of translating a digital model into thin stratification layers.

**Stereolithography (STL) file** is a file format native to the stereolithography CAD software created by 3D Systems. STL has several after-the-fact backronyms such as "Standard Triangle Language" and "Standard Tessellation Language". It is now a file format that holds 3D modeling and slicing information readable by a 3D printer.

## Slicing the 3D Model

After your object is in a usable file format, use slicing software to *slice* your object into printable layers and prepare your object for print. We will talk more about both free and purchasable slicing software in Chapter 7.

The slicing program—according to your settings—checks for errors, slices the 3D object into layers, adds *supports* where needed, and makes fill patterns for the interior. The 3D printer uses this information and slices to print the object layer-by-layer. **Figure 1.7** shows an example of a 3D object within a slicing program.

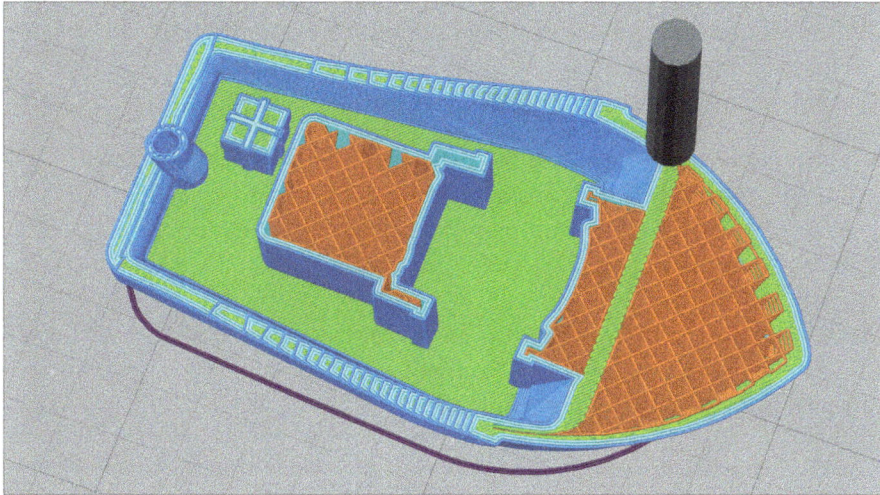

FIGURE 1.7 – A 3D model within the slicing software that shows the layers and *infill*. Jonathan Torta.

In the slicing program, software tests are run to check for errors in the design. The slicing program checks for problems such as overlapping sides, unaligned gaps, intersecting faces, or a *non-manifold* object.

For example, all 3D models are created by defining the surface of the object, and it is required that the object be *manifold* or watertight. Essentially, the 3D model is not solid. It is hollow and should be able to hold water if filled. A problem occurs if the object is not watertight and does not have all of the edges continuously connected. The slicer program may fail to correctly slice the object. Any errors can affect the quality of the slice. We will take a closer look at the types of errors and their causes in Part 2.

**BRIEF REVIEW OF TERMS:**

**Infill** is structures added for strength in enclosed empty or hollow areas in a 3D print.

**Manifold** is when 3D model geometry is "water-tight". Has a contiguous and defined exterior and interior.

**Non-manifold** is when a 3D model geometry is not contiguous like unwanted holes or has errors like intersecting edges.

**Supports** Additional printed structures that secure portions of printed objects that overhang greater than 45 degrees.

MAKER'S
NOTE

Various slicing applications have different parameters and may flag different errors. They may also indicate whether the errors are fixed automatically or manually. However, slicing application error-checking is not fool-proof: the errors found are very dependent on the error checking implementation—some slicers may have a more robust error-checking implementation, others may just flag the errors.

Slicing applications are not the only way to catch errors. Online validators and dedicated model editing applications are a few examples of other ways. Catching any errors before importing the 3D model into the slicing program is preferable.

Mechanical structures, like supports and *rafts*, may need to be created for the object to be printed correctly (see **Figure 1.8**). These supportive structures are removed from the object after printing.

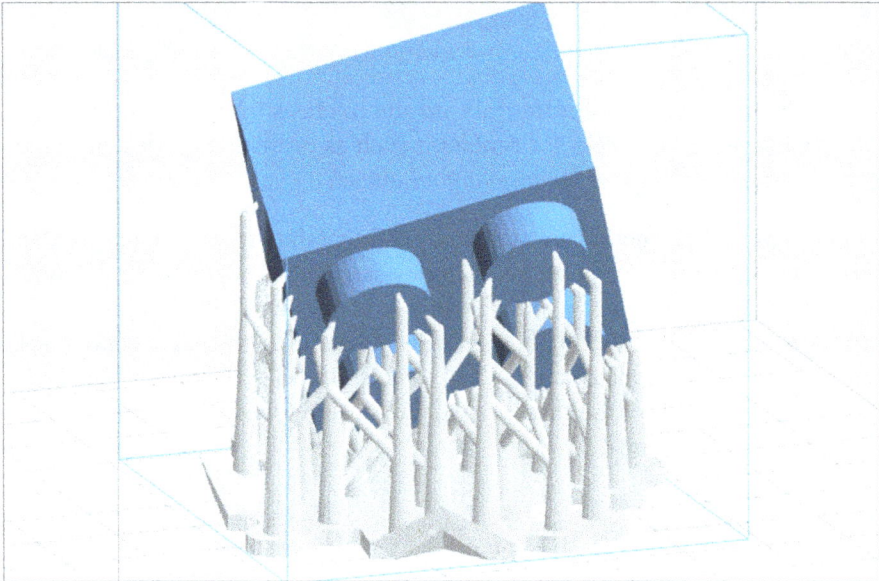

FIGURE 1.8 – The support structures for a LEGO® block. PranjalSingh IITDelhi.

The slicing software then slices and converts the design into thin layers as a blue-print, or *G-code,* that the 3D printer follows to create the print, layer by layer. This code carries all the instructions that the printer needs to print the object, like temperatures (*hot end* and *bed*), paths of each layer, and fan start and stop times.

MAKER'S
NOTE

The slicing programs can use additional proprietary codes other than generic G-code for the printer to follow.

**Bed** is the printing surface area where the 3D print is made.

**G-code** is a numeric control programming language that communicates the slicing information to the printer. This includes speed, location and path directions.

**Hot End** is the component where the filament is melted to a desired temperature.

## Printing

There are a variety of ways you can access a 3D printer. Some of the more common are buying or building your own printer like retail stores (like hardware and computer stores) and kits. Other ways to gain access to 3D printing is renting, public libraries, makerspaces, or online services.

> It is recommended to conduct research on types of printers or printing services to find one that will best fit your printing needs for size, material, timeframe, and quality.

MAKER'S
NOTE

After the slicing program is finished, the sliced model information is then communicated to the printer by connecting your computer using *Universal Serial Bus (USB)* cable, *Wi-Fi*, or *Secure Digital (SD) card*. See **Figure 1.9** for a

FIGURE 1.9 – A typical 3D printing setup, including a FELIX 3D Printer currently printing, a Macbook running 3D printing software, and some example 3D printed objects. Jonathan Juursema.

closer look at a connected laptop with an open slicing program with a 3D model being sent to the printer. Online services will have their own upload process.

Depending on the type of 3D printer, a variety of materials can be used, including plastics, metal, ceramics, glass, paper, and food (like chocolate). We will talk more about the different types of printable materials in Chapter 6.

The printer then prints the object; a print can take several hours. The 3D printer does exactly what computers and machines do best, methodically and precisely follow instructions—in this case tracing a line of plastic (or other material) layer-by-layer (see **Figure 1.10**). After the printer is finished laying down the layers of material to create the object, any support structures can be removed.

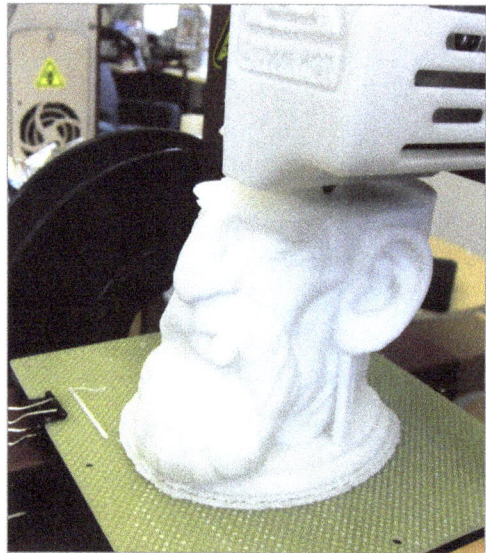

FIGURE 1.10 – 3D Printer Shootout Testing – Day 3. John Abella.

## Refining the Print

After the print is complete, some refinements may need to be done before the object is finished, like cleaning and removing any stray material. Because the printing process involves the bonding of material layers to create the object, the stratification will show in the print. There are several different methods to smooth or cover the layers. A few examples are sanding and painting. **Figure 1.11** shows a 3D printed replica that was finished for display. We will cover finishing in Chapter 13.

FIGURE 1.11 – A 3D printed tactile replica of the Tsar Cannon. Mos.ru.

## › 1.3 – What are some common parts used in a desktop 3D printer?

3D printers can be operated by a variety of printing methods. We will cover some of the more common ones in Chapter 5. In this book, we are showcasing a type of 3D printer known as *Fused Deposition Modeling (FDM)*.

An FDM printer is one of the more popular desktop 3D printers on the market today. They work by heating up *thermoplastic* filament to a desired melting point then extruding the heated material layer on top of material layer to form a 3D object. Or another way of thinking of it is as a computer-controlled hot glue gun that builds an object layer by layer. We will use this printer type to show a few of the more common parts and their function.

### BRIEF REVIEW OF TERMS

**Fused Deposition Modeling (FDM)** refers to the process of depositing continuous heated material in layers to create an object. Because this term is trademarked by Stratasys Inc., the term Fused Filament Fabrication (FFF) was created and can be used in place of FDM.

**Thermoplastic** is a type of plastic that can repeatedly become pliable when heated at a specific temperature and reverts to a solid state when cooled.

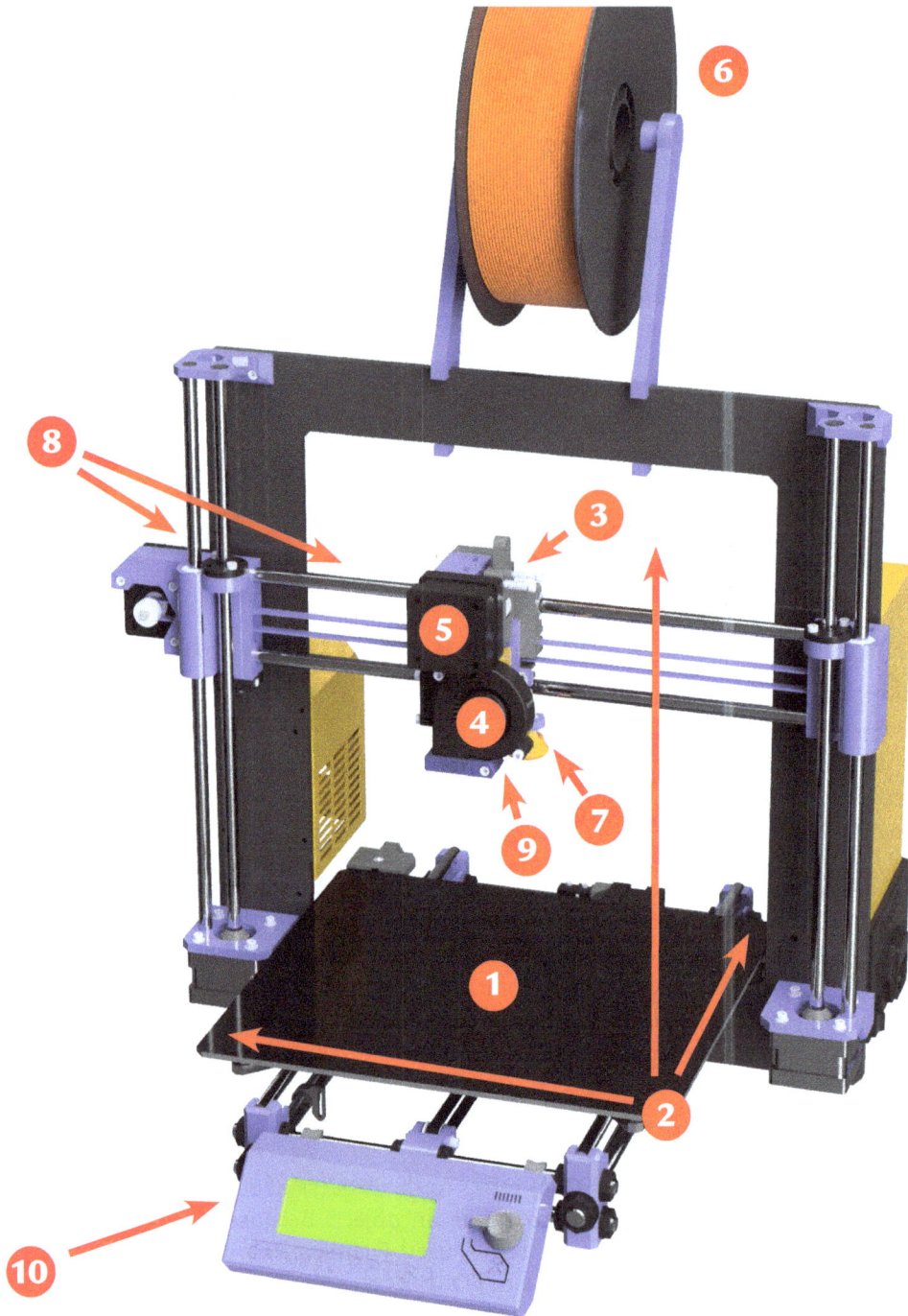

FIGURE 1.12 – A 3D printer with labeled parts. Jonathan Torta.

Depending on the type of 3D printer, the parts and their locations may differ from the diagram. See Chapter 5 for a more in-depth list of parts and their functions.

Understanding the parts of a 3D printer will help you gain knowledge in the inner workings of the printer along with those components you can adjust or modify to ensure the quality of your print (see **Figure 1.12**). If you buy or build your own printer, you will need to maintain it for consistent optimization.

The common parts of a 3D printer include:

1.  **Bed** – The printing surface area of the 3D printer. This is the component where the 3D print is deposited. It can be made of a variety of different materials including plastic, metal or glass.
2.  **Build area** – The overall size of the printable area that determines how large of an object can be printed. This includes the XYZ or width, height, and depth dimensions. Some build areas are cubical while others are cylindrical.
3.  **Cold end** – The part of the extruder where the filament is pulled from the spool then fed through or to the print head.
4.  **Cooling fan** – There can be multiple cooling fans located within a 3D printer. These fans help cool areas including the extruder motor, print head, and newly extruded filament for some types of materials.
5.  **Extruder** – or print head is a motorized device that has two assemblies: the cold end and the hot end. The cold end pulls filament and feeds it to the hot end that in turns heats the filament before the material exits through the nozzle into the build area.
6.  **Filament spool** – The mounting location for the spool of material for feeding into the extruder.
7.  **Hot end** – The part of the extruder where the filament is heated and melted to a desired temperature.
8.  **Linear rod or rails** – Used to help the print head (or bed) to move around reliably during printing.
9.  **Nozzle** – A small metal device with a fixed-size hole where heated plastic is sent through into the build area. Different size nozzles are used depending on the type of object being printed.
10. **Local controller** – A device that controls and gives commands to your printer without a computer.

## › 1.4 – How was the technology for 3D Printers invented and first used?

3D printing has become more common in everyday life and in pop culture in recent years. Because of this, it is often thought of as a new technology. However, the technology was invented and first used over 30 years ago. In the early 1980s, the first 3D printer capable of printing a physical part was invented by Charles (Chuck) Hull.[2] Other inventors built upon this groundbreaking technology to develop new machines, printer parts, and materials. The use of 3D printers started to expand and become more accessible.

The traditional manufacturing method of prototyping parts and tools was a long and laborious process, often using the subtractive method and carving out molds. Any changes to the design would often lead to starting the whole process again from the beginning. The use of additive manufacturing, including 3D printers, in the field greatly increased the speed and efficacy certain types of prototypes can be made or modified. The term *rapid prototyping* in this context refers to this method and spotlights the overall speed and effectiveness 3D printing brought to this section of industry. Designers could now test a tool or part design using a faster print-on-demand production method and use less expensive materials, allowing them increased flexibility to create or modify their designs.

With the continued advancement in 3D printer technology and materials, its use is expanding beyond rapid prototyping into the rapid manufacture of tools, parts, and products. This ability to effectively create multiple commercial end use products has opened the door for 3D printers in a vast number of industries and uses. We will illustrate some of these industries and uses in Chapter 2.

## › 1.5 – What is the potential for 3D printers?

With the cost of buying or building personal 3D printers decreasing, the availability of commercial 3D printing services, a wider array of materials, and the ever-expanding creative uses in industry, the potential for 3D printers' use is expanding.

When 3D printers were first created, they were used in specialized industries and were too expensive for small business or home use. This changed with advances in technology, like the advent of cheap microcontrollers and stepper motors, that lowered the overall cost and increased the demand. This allowed *Do It Yourself (DIY)* enthusiasts, schools, and small businesses to buy or build their own 3D printers (see **Figure 1.13**). The expansion and evolution of materials are also helping expand the growth and the creative uses of a 3D printer.

FIGURE 1.13 – A maker using a 3D printer. monkeybusiness.

## BRIEF REVIEW OF TERMS

**Do It Yourself (DIY)** is a phrase describing enthusiasts that build or modify objects without a professional craftsman.

With the flood of new users, the popularity of 3D printers grew, taking 3D printing from an intriguing novelty to a large presence in our pop culture. Television shows, movies, books, and video games all have featured 3D printers. For example, in the video game *Prey*, developed by Arkane Studios, the player used recycled material and a 3D printer to fabricate usable items.

Another example is the reboot of the television show *Lost in Space*, produced by Netflix. The crew uses a 3D printer in their spaceship to fabricate needed parts after crash-landing. The show *Cloak & Dagger* by ABC Signature Studios and Marvel Television showcased a 3D printer creating, layer-by-layer, a figurine of one of the protagonists. The printing scenes also showed supports being used and the finishing of the print. In the film *Ocean's Eight* by Warner Bros. Pictures, both a 3D scanner and a 3D printer were used in recreating jewelry.

While the growth of 3D printer use in the home or small business is becoming more commonplace, some of the real potential is still behind-the-scenes in industry. Some of the fascinating and rapid advances have been in the medical, dental, manufacturing, and aerospace fields. In Chapters 2 through 4, we will take a closer look into some of these fields along with how they are being used in education and in the home. We hear from Scott Higby for his thoughts on the potential of 3D printers in art classes.

**SCOTT HIGBY**                                         **QUOTE**

**HS/MS ART TEACHER, AFTON CENTRAL SCHOOL**

I think they [3D printers] would have great potential for any graphic design or sculpture classes that I teach. I had one of my advanced art students do a 3D project where he designed some engine parts and printed them on a 3D printer. I think students, if given the opportunity, would love to learn more about 3D printers.

## SUMMARY

The technology for 3D printers was invented over 30 years ago but has become more common in recent years because of the advent of inexpensive microcontrollers, stepper motors, and advances in technology. The potential of using 3D printers has grown and the cost has lowered. Unlike a flat two-dimensional (2D) printer, a three-dimensional (3D) printer uses length, width, and depth to make objects by using the additive manufacturing process of adding layer upon layer of materials. There are many steps in the 3D printing process, such as obtaining a 3D model, selecting a file format, slicing the 3D model, printing, and refining the print. There are at least nine common parts of a 3D printer (bed, build area, cold end, cooling fan, extruder, filament spool, hot end, linear rod or rails, and nozzle). The next chapters will present in-depth information about 3D printer use in society.

## APPLYING WHAT YOU'VE LEARNED

1. Start your own 3D dictionary by adding the definition (in your own words) of five words relating to 3D printing that you did not know before.

2. In the future, what do you think 3D printers will be able to do and why?

3. Describe a project you would like to make with a 3D printer and why would you like to make it.

4. What is the first process you must do when starting to make a 3D print and where would you find the information?

5. Explain what a 2D printer and a 3D printer are and how they are different and the same.

6. Explain one of the greatest advantages of using a 3D printer and give some examples.

7. Briefly describe the process of the flow chart in this chapter in your own words.

8. Describe eight parts of a 3D printer and how they are used.

9. What manufacturing process is 3D printing a subset of? Describe this manufacturing process?

10. In your own words, how are the terms CAD, G-Code, modeling, raft, slicing, and supports used in 3D printing?

11. Describe the three manufacturing processes and how are they different from each other.

12. Draw your own picture of a 3D printer and label each part.

## REFERENCES

[1] – http://additivemanufacturing.com/basics/

[2] – https://www.3dsystems.com/our-story

# CHAPTER 2

# 3D Printer Use in Industry

## OVERVIEW AND LEARNING OBJECTIVES

**In this chapter:**

- 2.1 – How are 3D printers being used in industry?
- 2.2 – What is the potential growth of 3D printer use in industry?

Although 3D printers were first utilized in industry, they have expanded almost everywhere, including small businesses, education, and consumer markets. In this chapter we will highlight a few ways 3D printers are being used in a cross-section of industries and businesses such as:

- Aerospace
- Archeology
- Architecture
- Culinary arts
- Dentistry
- Entertainment
- Medical

We will also talk with several makers to hear their firsthand accounts on how 3D printers are incorporated into their occupations.

**Web search**

What fields are you interested in? Do a web search on how 3D printers are changing or enhancing those industries or hobbies.

**IN EXTRAS**

*This chapter notes several web URLs. These links are also in the extras and on the DVD in an interactive PDF for quick viewing.*

## › 2.1 – How are 3D printers being used in industry?

Industries ranging from automobiles to architecture, to art, to fashion all use 3D printers. Because each field has specific production requirements, 3D printers have evolved to fit the needs of a particular industry by inventing new printing processes and developing new materials.

**Web link**

The link below is to a PBS news segment talking about evolving 3D printers by developing new custom printers and print methods to decrease production time and costs while maintaining the quality of the product.

https://youtu.be/Adl1Sn86ojs

3D printers are changing production and workflows within industry in many ways. In **Figure 2.1** we see a mechanical part being printed as an example of rapid prototyping.

It is important to understand that 3D printers come in all shapes, sizes, and use a variety of printing techniques and materials. This range includes large printers that can print houses, to printers using robot arms, to printers that use lasers

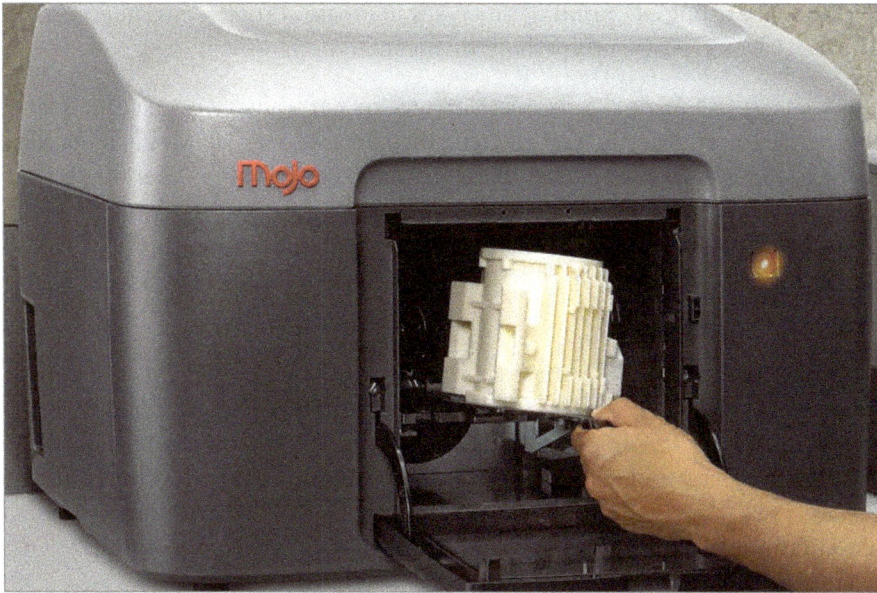

**FIGURE 2.1** – Mojo 3D Printer. Intel Free Press.

during the printing process. Most industrial printers are not the desktop types you would find in schools, libraries, or the home. They often are specialized and use more complex printing processes. In Chapter 5, we will discuss a few types of 3D printers and showcase how they may differ in the printing process.

For this section, we are highlighting a few industrial fields to show examples of the wide range of uses and how 3D printers are changing our world.

## Aerospace

The aerospace industry is a driving force in innovation and their use of 3D printers is no exception. The 3D printers' ability to prototype rapidly is a natural fit for testing aircraft and spaceship design. However, their use has expanded beyond concepts into manufacturing machine parts and printing on-demand in space.

The National Aeronautics and Space Administration (NASA) is utilizing 3D printers both in space and on the ground and is driving the evolution of additive manufacturing and 3D printers.

## Spacecraft machined parts

When you think of 3D printing, you may not think of the ability to print rocket parts for spacecraft; however, with the advances in printing technology and materials, they are able to do just that. NASA facilities have worked together to combine existing 3D printing processes to create complex mechanical parts. This has lowered costs and sped up the manufacturing timeline.

One example is NASA's development of a rocket part called a pogo accumulator. Being able to 3D print the part reduced the need for over 100 weld points. This reduced the overall cost by over 30 percent and build time by over 80 percent. The part was successfully tested and has led the way for further development in printing rocket components.[1] Another example of time savings is the 3D printing of large titanium satellite fuel tanks by Lockheed Martin. Production time was cut from two years to three months.[2]

NASA also created new 3D printing processes to fit specific needs, most notably to maintain quality in extreme temperatures and pressures. An example is the printing of rocket engine nozzles. These nozzles are complex parts, subject to extreme heat and cold. They have channels within the structure that must be precisely sealed to contain the high-pressure coolant. A new wire-based method of 3D printing was developed called the Laser Wire Direct Closeout (LWDC) technique. Using this process, NASA was able to print the nozzle, close the coolant channels, and create supports to help with structural loads during operation. The nozzles have been successfully tested and now have the potential to reduce production time from months to weeks.[3]

In **Figure 2.2**, we see an example of a 3D printed combustion chamber rocket part being tested for quality. This part was printed using a combination of different printing techniques that reduced the overall number of parts to increase durability and decrease production from weeks to hours. [4]

**Web link**

**See the testing in action:**
**Low Cost Upper Stage-Class Propulsion project (LCUSP)**

**https://youtu.be/FfeP4BVTsps**

## Print on-demand in space

Space travel is like a very long hiking trip into the wilderness. Whatever items you will need during the trip, you must carry with you. Cargo space and weight become issues, along with deciding what you should bring to cover any unexpected occurrences during the trip. The more you bring, the better prepared you are, but the more difficult it is to carry everything. Depending on where you hike, your survival could depend on your decisions. For space travel, you have the added issues of carrying the power needed to escape our atmosphere or carrying all the materials needed to start a surface-side base.

FIGURE 2.2 – NASA successfully hot-fire tested a 3D printed copper combustion chamber liner with an E-Beam Free Form Fabrication manufactured nickel-alloy jacket. The hardware must withstand extreme hot and cold temperatures inside the engine as extremely cold propellants are heated up and burned for propulsion. NASA/MSFC/David Olive.

A goal for NASA was to develop a way to work around some of these issues. They looked to 3D printers and the ability to print on-demand. The ability to print parts and tools when needed decreases the number of things you need to launch with you and gives you flexibility with unknown incidents. While this might help solve the issue of what to bring, a few questions are still to be answered.

- Will a 3D printer work in zero gravity?
- Is there a way we can use local materials to print rather than bringing material?

To solve the first of these questions, NASA and its partners developed a custom 3D printer and installed it on the international space station. The tests were successful, showing that 3D prints were able to print tools and other objects in space.[5] [6]

Solving the amount of mass carried during launch was the next issue. NASA hopes to eliminate the need for structural materials to be launched from Earth. As part of the Zero Launch Mass goal, researchers simulated planetary resources and invented a system for space travelers to 3D print construction projects using those materials. **Figure 2.3** shows a print on-demand structure made with materials that are like what can be found on destination locations like the Moon or Mars.[7] [8]

FIGURE 2.3 – The Zero Launch 3D printer. NASA/Glenn Benson.

**Web link**

**Read more about NASA and their evolving use of 3D printers:**

**https://www.nasa.gov/topics/technology/manufacturing-materials-3d/index.html**

These new printing techniques and technologies are not only used to manufacture rockets; they can be used across industry to create mechanical parts for complex machines such as airplanes and cars. These breakthroughs will potentially have far-reaching and direct impact on manufacturing production workflows, changing a multitude of industries.

# Archeology

The study of our past is not something that comes to mind when thinking about 3D printers. However, 3D printers and 3D scanners are often used together to help us understand our history by recreating, fixing, or reimagining relics and ruins of our past through non-invasive methods.

## Non-destructive research

3D scanning and printing have several non-destructive benefits to museums and researchers. For example, one of the biggest problems in studying artifacts or ruins can be their singular location, rarity, or frailty. Researchers often must travel to the artifacts or ruins. This can be difficult and time consuming. With the use of 3D scanners, detailed 3D models can be created using a non-destructive process, preserving the integrity of the artifact. These 3D models can be used to print high-quality reproductions for study. This allows exact replicas of the valuable and delicate artifacts or ruins to be sent to the researchers around the world at any time.

In **Figure 2.4** we see a 3D scanner being used to create a 3D model of an artifact that was later reproduced using a 3D printer.

FIGURE 2.4 – Making a 3D model of a Viking belt buckle using a handheld VIUscan 3D laser scanner. As the device uses a laser scanner to create a 3D model, it also uses a camera to accurately texture map the object. This 3D file is the result of a project in which a 1000-year-old Viking belt buckle was laser scanned and 3D printed to achieve a copy of the unique archaeological artifact. Creative Tools.

These 3D scanned models typically carry more detail than a common 2D photograph. When printed, researchers have the added benefit of being able to hold a tangible exact replica of the artifact for closer study rather than relying on a series of photographs.[9]

Some artifacts can be damaged through time. To preserve artifacts in their current state, 3D scanners can be used to capture detailed 3D models to archive entire collections and store them digitally for future comparisons or reproductions.[10]

Restoring or repairing artifacts is another way 3D scans and prints are valuable. Missing parts or gaps can be reproduced for an exact patch to complete an artifact. Skeletons can be 3D scanned and printed to rearticulate the bones without damaging the originals.

## Hands-on reproductions

Connecting to our past can be enhanced when we can physically touch and hold an artifact from our history. 3D printers and scanners have enabled scientists and the general public to interact with historical artifacts and ruins in a way never previously possible.

## Reproducing

One example of hands-on history is reproducing ancient jewelry using the original molds. In the Orkney Islands in northern Scotland, Iron Age clay jewelry molds were discovered. One of these molds was used over 2,000 years ago to create a bronze ring-headed pin. With the help of 3D scanning and printing technology, a non-destructive method was used to produce a new bronze pin following the steps the original creators took.

To achieve this, the clay mold was photographed and scanned to create a detailed 3D model. This 3D model was then inverted to create a new model of the pin itself. This 3D model was then printed in wax. Afterwards, the 3D printed wax pin was pressed into clay to make a new clay mold. Bronze was poured into the new mold, melting the wax pin in the process. Once cooled, the clay mold was removed, revealing a new bronze ring-headed pin inside.[11]

This process not only allows researchers to get a glimpse of the production processes of the past, but also to hold an exact replica of the artifact, seeing in detail the craftsmanship up close. With 3D models, detailed prints and reproductions of the models and pins can be made for both scientists and the public without harming the original clay molds, bringing to life an artifact from the past.

## Availability

Another example of hands-on history is making museum and other artifact collections more available to people by using 3D scanning and printing. The 3D models and reproductions can bring the artifacts to the general public both within the museum or institution and within their own home.

While visiting a museum or institution, 3D scanned, modeled, and printed artifacts can:

- Engage visitors by letting them touch, hold, learn, and inspect priceless artifacts without damaging the original
- Be sent to other museums and locations easily to increase the availability of the artifacts so that they can be viewed by more people
- Allow visually impaired people to connect with the artifacts by printing tangible replicas they can interact with
- Produce reproductions for visitors to buy from museum gift shops

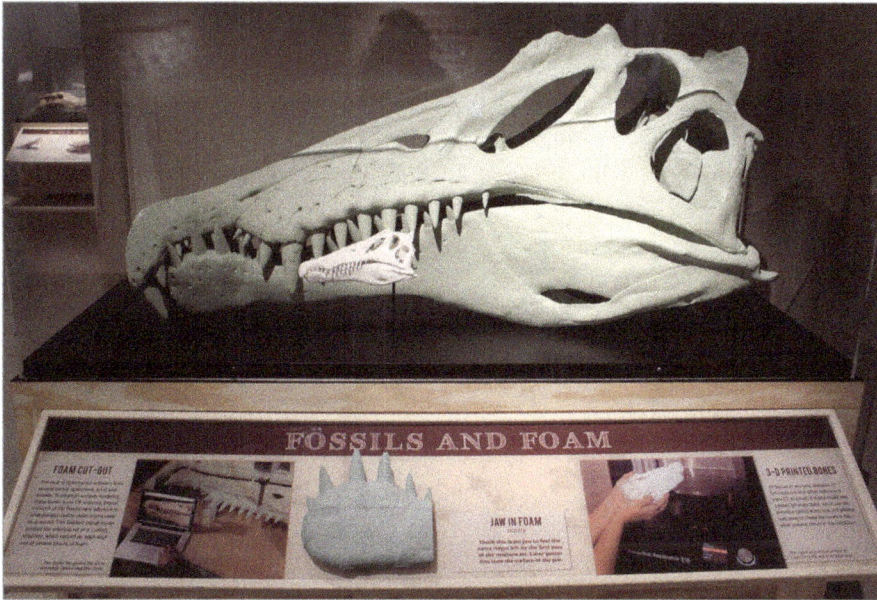

FIGURE 2.5 – Taken at the National Geographic Museum Spinosaurus Exhibit. Ryan Somma.

In **Figure 2.5** we see a museum exhibit showcasing a Spinosaurus skull. This skull was composited from multiple scanned partial specimens. The new 3D model of a complete skull can now be cut from foam to create large exhibit pieces or 3D printed on a smaller scale for audience interaction.

3D technology is also helping museums and institutions expand beyond their walls. Some museums, like the British Museum and the Smithsonian, have created 3D models of artifacts from their collections. These models are listed online as virtual tours and can be downloaded. They can be viewed and rotated in a virtual space or 3D printed. This allows individuals, libraries, and schools access to their collection anytime or anywhere. [12] [13]

> **View or download 3D models from the British Museum and the Smithsonian:**
>
> **https://sketchfab.com/britishmuseum**
>
> **https://3d.si.edu/**

Web link

We are only scratching the surface of how 3D scanners and printers can be used in archeology. 3D technology will keep evolving in the future to help us preserve our history.

## Architecture

As with most industries, the use of additive production in architecture covers a wide array of uses, ranging from planning to the building stages and from small to large structures. For example, **Figure 2.6** shows a printed scaled model for architectural planning, while **Figure 2.7** shows a completed 3D printed bicycle bridge.

FIGURE 2.6 – Architectural model printed with an Ultimaker 3D printer. Svitlana Lozova.

FIGURE 2.7 – The world's first 3D printed bridge. Marczoutendijk.

One of the primary advantages 3D printer use has in architecture is bringing to life 2D drawings of structures. In the planning stages, a 3D printed model helps the client visualize the architect's vision. An artist can now print complex scaled models of their structures quicker and less expensively than using past methods. What might have taken some artist months can now take hours, with improved quality. This helps both the artist and the client see what works and what still needs work in the planning stage. It also allows quick implementation of updates and variations.

A more complex advantage 3D printers are bringing to architecture is the building of actual usable structures. This ability is still relatively new, but the potential will drive it forward, inventing new printers' processes and materials. [14]

Some potential advantages 3D printing usable structures are:

- **Less waste** – materials can be optimally used
- **Creative design** – able to build freeform curves and other shapes effectively
- **Increased structural supports** – with freedom of design, structures can be printed with maximum structural support
- **Savings** – in materials, time, and money

There is still a long way to go before we see 3D printers at construction sites; however, the first steps have been successfully taken.

## Building on Demand (BOD)

An example of 3D printers being used to build structures is the Building on Demand (BOD) project in Copenhagen, Denmark, by 3D Printhuset. One of the project's goals was to build Europe's first 3D printed building while adhering to strict European building codes. 3D Printhuset tested concrete mixtures for optimal use in 3D construction printers.

Curves and organic shapes are difficult to build and cost-prohibitive, but a 3D printer can effectively print these shapes. To showcase this advantage, the BOD building has no straight elements other than the windows and doors. Even the roof is sloped.

**Figures 2.8 to 2.11** show the BOD building project through stages of production and the 3D printer in action.

Although the building was small, it was large enough to see cost savings and architectural freedom during production. The environmental impact was less than traditional building methods, not only because of the reduction in waste,

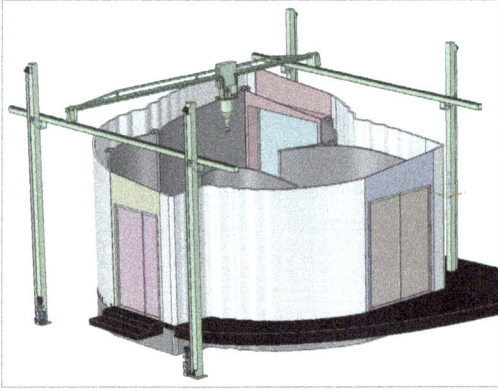

FIGURE 2.8 – A model of the building on demand (BOD) with the printer. 3DPrinthuset.

FIGURE 2.9 – The building on demand (BOD) printer developed by 3DPrinthuset. 3DPrinthuset.

FIGURE 2.10 – The finished 3D printed walls of the building on demand (BOD) building. 3DPrinthuset.

FIGURE 2.11 – The building-on-demand (BOD) structure is a recently built small office hotel with a 3D-printed wall and foundation structure. This is the first building of its kind in Europe. 3DPrinthuset.

but also because recycled materials were used as part of the concrete mix. The BOD project was able to show that 3D printed structures can be built to high European building code standards while reducing build time, waste, and cost.

**Web link**

**Read more about the 3D printing building-on-demand (BOD) building project:**

https://3dprinthuset.dk/europes-first-3d-printed-building/

The benefits of 3D printing to architecture are still being expanded and invented. The amount of money and time saved, environmental benefits, and creativity are going to continue to grow.

# Culinary Arts

In the culinary industry, specialized 3D printers are preparing food in both restaurants and the home. While 3D printers will not be taking over kitchens, they can be a useful and creative tool in food preparation. Like using other kitchen appliances, 3D printers have unique features that can assist in food preparation in conjunction with other appliances or on their own (see **Figure 2.12**). We hear from Lynette Kucsma, co-founder of Natural Machines for her insight.

FIGURE 2.12 – Foodini 3D food printer being used in a restaurant. Natural Machines.

---

**LYNETTE KUCSMA**                                                        **QUOTE**

**CO-FOUNDER & CMO AT NATURAL MACHINES**

Once you think about a 3D food printer as one of your kitchen appliances that can be used to create a dish, perhaps with other kitchen appliances helping out, then the variety of dishes that you can imagine being created with a 3D food printer grows tremendously.

---

As with other industries, the 3D printer's ability to rapidly manufacture and automate is a key feature they bring into the kitchen. Complex shapes can be identically printed, freeing the chef to work on other tasks during the printing stage (see **Figure 2.13**). For example, in preparation for an event where several dishes are all going to be the same, a chef can print repetitive or complex parts of the dishes that would normally take a lot of time. This chef is now free to work on other parts of the preparation.

FIGURE 2.13 – Guacamole dish. Natural Machines.

Another example is making foods that must be shaped or layered, like crackers, cookies, and ravioli. These can be programmed into a 3D printer and set to be shaped or layered on its own (see **Figure 2.14**). 3D printers can be an effective timesaving tool in the overall workflow.

FIGURE 2.14 – Hummus Castle. Natural Machines.

The type of ingredients that can be used in 3D food printing is widening. In the past, printing was limited to foods that had a thin paste consistency and ingredients that could be heated and cooled, like chocolate.

Now, almost any ingredient that has a certain degree of consistency after being prepared by hand or processed in a blender or food processor can be used. Even ingredients with a range of textures and whole chunks like walnuts and cranberries can be used.[16]

One example is the Foodini 3D food printer. It has different size nozzles that allow for a wider range of ingredients (see **Figure 2.15**). These food capsules can be filled with ingredients you prepared yourself. This allows you to print fresh foods such as pastas, burgers (meat or vegetarian), quiche, baked goods (crackers, breads, and cookies), or sweets. This is only a short list of potential foods you could print. Creativity in the kitchen extends to the use of 3D printers.

FIGURE 2.15 – Natural Machines Foodini stainless steel food capsules. Natural Machines.

Reducing waste from ingredients is another feature of using 3D printers in food preparation. For example, some people do not like the appearance or texture of certain foods. Some people only eat a part of the food, leaving the rest to waste. One such food is broccoli. In **Figure 2.16**, we see a 3D printed star made from chopped broccoli. Both the florets and the stalk were used in the star. This changes the look and texture of the broccoli and uses all edible parts in the printing.

This also has the side benefit of healthier eating because the food can be prepared fresh. Fresh ingredients can be prepared in batches then printed when ready for serving.

FIGURE 2.16 – A. Cut broccoli. Envato – sabinoparente   B. Natural Machines star broccoli. Natural Machines.

**Web link**

To learn more and to see the Foodini 3D food printer by Natural Machines in action, visit:

https://www.naturalmachines.com/

3D printer use in the culinary industry is still evolving. For a closer look into 3D printers' use in the restaurants and the kitchen, we talk with Liam MacLeod about his experience and research in exploring the 3D technology potential.

## INTERVIEW

### LIAM MACLEOD

*(Formerly) 3D Printing Specialist, The Culinary Institute of America*

**ST: How do 3D printers fit into your business?**

**LM:** The Culinary Institute of America was looking to explore the possibilities of utilizing 3D printing technology as an innovative skill in the provision of a unique sensory experience for a guest or customer. At the time, the college was exclusively in this exploration phase and had not monetized any products or formulations during my tenure.

**ST: What benefits do you find in using 3D printers as part of your business process?**

**LM:** As previously stated, The CIA was not in the production phase, merely exploratory. However, the outlook on the technology was as such: innovation in an industry as old as food is extremely sparse in this day and age. An unbelievable amount of processes and techniques have been discovered and explored, therefore it has become rather difficult to provide new experience for guests and customers. By utilizing printing technology, we could allow the guest to experience something new, garnering wonder and excitement, and in turn, develop repeat customers who will pay premium prices for these products.

**ST: What type of printer(s) or printing service do you use?**

**LM:** During my tenure, I utilized Thermal Inkjet (TIJ) processes in the form of the 3D Systems Zprinter 310 and ZPrinter 510 machines. These machines were early prototypes of the Chefjet Pro: a TIJ printer designed for use in a professional kitchen. As such, they were custom-modified and fitted with stainless steel components to allow for easy sanitation and food safety.

Currently, the college is exploring the possibilities of using Fused Filament Fabrication (FFF) machines in the form of the Cocojet, co-developed by Hershey and 3D Systems.

**ST: What software and material do/did you use?**

**LM:** For interaction with the printers themselves, I used the proprietary ZPrint Software developed by 3D Systems. Contingent upon the nature of the project, I would use one of two CAD software's to develop a 3D Structure. Primarily, Dassault Solidworks was my program of choice. However, occasionally the project warranted a more "physical touch," in which case, Geomagic Freeform paired with a haptic device was the ideal software.

As far as material is concerned, I am not at the liberty to disclose exact formulas or constituents. However, I can say that I utilized a complex blend of carbohydrates, fats, and proteins as a powder base, and a water-based blend of various surfactants, parabens, preservatives, and color as a fluid binder.

For the FFF printing, the college is currently using a modified chocolate product, optimized for its melting plasticity.

**ST: How did the use of 3D printers affect your production timeline?**

**LM:** Again, we had not formally produced using the machines at the time. On the surface, the process did not seem to be efficient and fast-paced enough to match a hectic kitchen environment; the printing process of layering thousands of times combined with the curing stage was certainly not quick. However, considering the products we developed were meant merely to accent a chef's creativity, not outshine it, the desired product was quite small. Because of this, we could print upwards of 100 items per load. The machines could run autonomously, and even overnight with minimal or no supervision—push "print" and go home. In turn, we found that it was possible that, for minimum effort (outside of 3D modeling), a chef can produce hundreds of products comprised of inexpensive material, which patrons would pay premium pricing for each day.

**ST: What iterations did you go through?**

**LM:** I would follow a standard R&D iteration pipeline: hypothesis development, market research/focus group, concept development, prototype development/optimization, and product testing.

The most substantial step in my operation was certainly product optimization. We were provided with a few basic formulas which were developed to mimic the physical properties of polymer powders (as these machines were originally designed for that material). While certainly edible, these formulas did not provide a flavor experience commonly expected in a fine dining format. For us as culinarians, if the item does not taste excellent, it is useless. Therefore, I spent a lot of time optimizing the flavor and textural components of both the powder and the binder by incorporating various constituents in certain percentages. The difficulty then became, again, matching the properties of the old polymers through analysis of powder and binder rheology, brix and pH of the binder, powder hygroscopicity, porosity of the powder feed reservoir, and other considerations as well.

**ST: What did you change during the process?**

**LM:** It is important to remember that, while 3D printing is not a new process by any means, there are very few documented instances of using the technology to create edible products. It was difficult to foresee the outcome of each product, and because of this, basic formulas for powder and binder were constantly being changed to optimize flavor, structural stability, melting profile, financial consideration, time optimization, and more.

**ST: Was there a cost savings?**

**LM:** Because there was no production component, a cost savings could not be calculated. However, we can anticipate that guests would pay a premium price for a new experience based on the financial success of innovations in the past.

**ST: How did you decide to use a 3D printer for the project?**

**LM:** A few years back, representatives from The Culinary Institute of America and 3D Systems met at a national conference focused on advancements in food technology. 3D Systems was presenting their patented "ChefJet Pro": a printer designed for use exclusively in a professional kitchen. Their machine was capable of creating the complex 3D structures out of edible products rather than polymers. As an institution perpetually at the forefront of advancements in the food industry, The CIA was intrigued by the capability of the printer, discussed collaboration between their Chefs and 3D Systems engineers, and developed a partnership.

3D Systems provided the equipment necessary to operate a laboratory within The CIA, and The CIA provided expertise on the organoleptic experience of the finished products, advancements and improvements on flavor and texture, and their functionality on a plate. The lab was started by a CIA student who had previous experience with the equipment as an engineer, and I took the reins soon after.

**ST: Did you start with a physical model, a drawing or a CAD design?**

**LM:** Using both Solidworks and Geomagic Freeform (with a haptic device), we developed our own 3D Structures in-house. These models began as thoughts taken straight out of a chef's mind, drawn for approval, translated into 3D, and printed for testing.

**ST: What do you see as the future of 3D printers in your business?**

**LM:** Food preparation has been a critical component of evolution throughout human history. From the moment Homo erectus controlled fire onward, we as a species have been advancing and innovating with our food at a logarithmic. Fast forward to a more modern era, and our viewpoint of eating has been turned upside down: it is not seen exclusively as a survival tactic, rather, a recreational experience to be enjoyed. We have built industries behind food, food preparation, manufacturing, and are even fortunate enough to have the means to develop a preference over what food we would like to eat any given day.

Because of this, we get bored of eating the same things prepared the same way over and over again. We look for inspiration; we look for new experiences and in an industry whose roots were planted more than half a million years ago, it is unbelievably difficult to innovate beyond what has already been achieved. While printing technology is not a new process by any means, its functionality in a kitchen has been only lightly explored with very little results. We see the future of this technology as another tool which a chef may keep in their back pocket to provide their guests with a sensory experience they have yet to explore.

**ST: Are there any project or stories you would like to share in utilizing 3D printing?**

**LM:** In my experience, I have found that when someone is to incorporate a new technology into a specific field (particularly one built on the idea of skilled craftsmanship), there are generally only 2 reactions by their peers. There are those who see the new tool as an extreme convenience, maximizing their output and in some cases creativity, and there are those who see it as a mechanized menace threatening to replace the human hand.

I was met with both sides of that coin at The CIA and outward.

It is important to remember that you would be hard-pressed to find a more skilled and qualified congregation of craft masters than you would at The CIA. There is a reason it is "The World's Premier Culinary College." These are individuals who have not only experienced, but performed at the highest level of their trade for decades upon decades. Needless to say, it's a tough crowd to impress. Those who were interested loved the new tech. They loved the visual perfection of the product and the new windows of creativity it opened for them. Those who hated it simply did not understand the mission. We were not looking to replace a chef with a machine, and this was exemplified by all of the projects performed. Our goal was to simply accent a chef's craft by creating a small, bespoke component to add to a dish that could not be replicated by hand: a garnish, or vessel, if you will. In other words, the chef's work will always be "the star of the show," and the printers were merely a tool in their wheelhouse to create something new, exciting, and above all, fun.

The use of 3D food printers in the culinary industry is still at the beginning stage, but, is growing. 3D food printers can bring a creative automation to the kitchen as useful time and ingredients saving tool (see **Figure 2.17**). In the future, 3D printers could be a common appliance in kitchens in both restaurants and the home.

FIGURE 2.17 – ChefJet Candy 3D Printer. Maurizio Pesce.

## Dentistry

One of the fastest growing areas where 3D printers are being used is in the dental industry. The 3D scanner's and printer's ability to quickly customize crowns, dentures, dental implants, and braces has revolutionized production processes (see **Figure 2.18**). Additionally, the ability to make the objects dentists need locally and on-demand decreases the turnaround time for many procedures and allows additional custom work. The dental industry is currently one of the largest buyers of *Stereolithography (SL) SLA* and *Digital light processing (DLP)* resin printers. Designated dental 3D scanners and printers are being produced to fit the needs and demands. [17, 18]

---

**BRIEF REVIEW OF TERMS**

**Stereolithography (SL)**, also known as Stereolithography Apparatus (SLA) is a registered trademark of 3D Systems. SLA uses an ultraviolet laser and mirrors to cure the resin.

**Digital light processing (DLP)** – DLP uses a projector to cure large areas of resin.

---

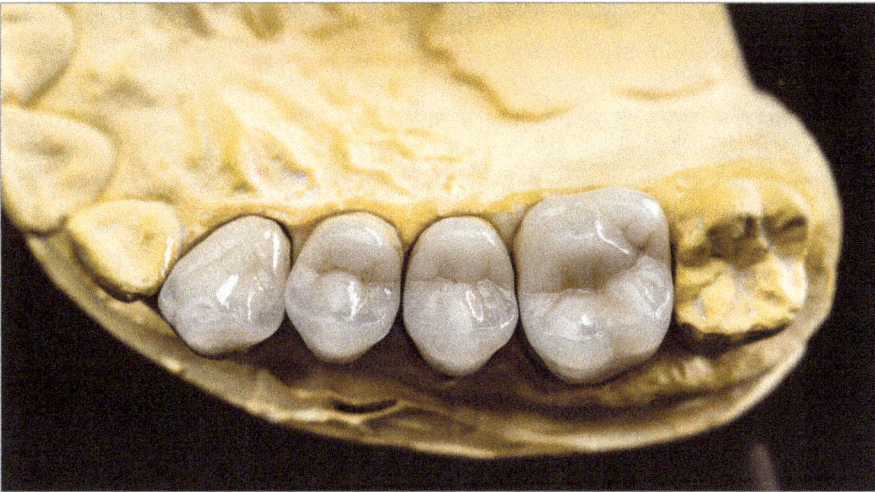

FIGURE 2.18 – An example of dental work being produced. Envato – grafvision.

MAKER'S
NOTE

We will talk more about SLP and DLP 3D printers in Chapter 5.

Each dental patient is different, with unique needs. For work that needs to be done, molds were taken of the patient's mouth. The mold was sent to a dental lab to create the needed dental implant. Once created, the implant was fitted to the patient. If there were any problem or fitting errors, the whole process had to be redone. This added time and money to the process.

With the introduction of 3D scanners and printers, this process is quickly changing. 3D scanners can now scan the mouth or molds of the mouth to create accurate 3D models. These 3D models can be used by the dentist, orthodontist, or dental lab to 3D print and fabricate the dental implant. If there are any changes that need to be made the dentist, orthodontist, or dental lab can tweak the 3D model without having to start the process over, saving both time and money.

For example, traditionally, if you went to the dentist for a night guard to help with grinding your teeth, you would have to bite into a thick paste to make an impression of your teeth to make a mold. This mold was then sent to a lab to create the night guard. Then it is sent back for fitting.

With 3D printers and scanners, the dentist can take a 3D scan of your mouth to create a 3D model while you are in the office. Then they use that 3D model of the mouth to design a new 3D model for the custom night guard. This new 3D model is sent to a 3D printer in the office or to a lab for printing. This process could take under an hour, compared to weeks. Also, if the guard does not fit correctly, adjustments can be made to the model and the night guard can just be quickly printed again (see **Figure 2.19**).

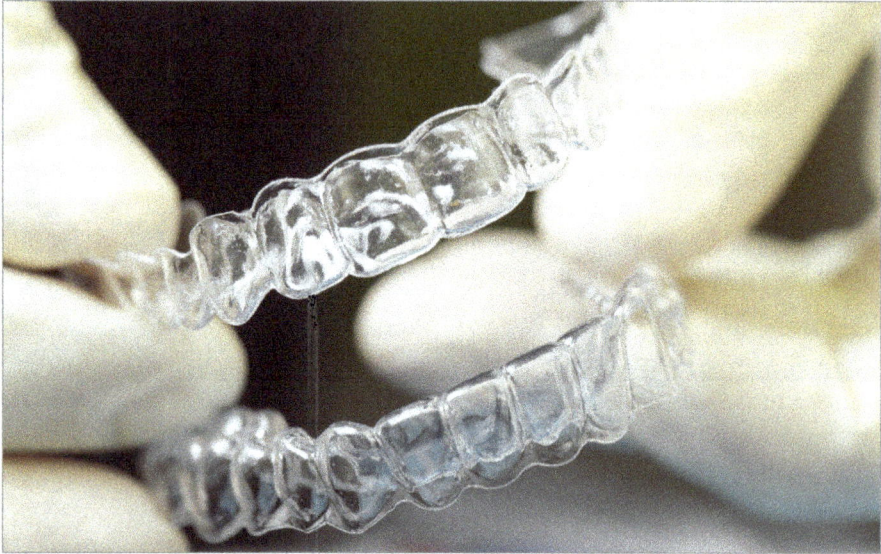

FIGURE 2.19 – An example of a dental mouthguard. Envato – Grey_Coast_Media.

A dental digital workflow can save space as well; 3D scans and models can be stored on computer storage devices rather than storing traditional physical molds.[19]

Lower costs and quick customization and turnaround times for each patient makes the use of 3D scanners and printers a perfect fit for the dental industry.

## Entertainment

The entertainment industry is a very broad and includes many different sub-categories. For this book we will focus on the performing arts, which include cinema, television, and theater.

At first, it's easy to overlook how effective the use of 3D scanners and printers can be in the production of movies, shows, or plays. However, a closer look will show how 3D scanners and printers have revolutionized types of movie prop construction and resource availability.

Traditionally, production teams had to find and adjust existing items for props and models to fit their needs or fabricate new objects from scratch using traditional methods. Both processes could take a long time and be costly. In addition, when the talent had to be fitted with custom masks or other wearable props, a physical mold was cast made of the actor. This process can be time extensive and uncomfortable for the talent.

With the use of 3D scanners and printers, the workflows for making props are changing. As with other industries, the main advantages 3D scanning and printing bring to making props are time savings, customization, and less expense than traditional methods.[20]

Artists and sculptors can now scan their work with a 3D scanner to create a 3D model. They can also design 3D models of props using 3D editing software. Once these 3D models are made, 3D prints of the props can be fabricated using a variety of materials and sizes to fit the needs of the production.

For example, the British television show *Red Dwarf* used a physical model of a spaceship named *Starbug*. During the series, this spaceship model was scripted to be shot from multiple angles and even crash. Because crashing could damage the model of the *Starbug*, a 3D printed copy was made.

To make a copy from the original physical model, a 3D scan was used to create a digital 3D model. After this was made, the new *Starbug* was 3D printed. Because they were now starting from a 3D model, they could print the *Starbug* in parts or in full at different scales depending on the needs of the script. After finishing the new 3D print by cleaning and painting, it was ready for the action shots.

During the various crash takes, the 3D model was damaged. However, it was an easy fix to print replacements for the damaged parts. A process that could take months was now done in a week's time. This allowed the production team to shoot until they got the right shots without having to worry about damaging the model.[21]

> Watch a video of the Starbug being printed and an interview of the process:
>
> https://youtu.be/Cnn2Zs3hDTc

**Web link**

Another example of 3D scanning and printing making the prop creation workflow more efficient was the making of Hela's headdress from the Marvel movie *Thor: Ragnarok*. The headdress was smooth, with large, antler-like horns extending from the base skullcap. The process started by 3D scanning the actress' head to create a 3D model and cast. This was used to create a custom fit for the skullcap. Next, the horns were digitally designed using 3D modeling software and 3D printed. Because the horns were large and weight would be an issue, a composite with carbon fiber was used. With the workflow being completely digital, the customization and alterations could be made to the 3D model and reprinted. After the design iterations were complete and a final was printed and assembled, the headdress was finished and painted.

During the final cut of the movie, shots of the headdress included a combination of computer-generated effects (CGI) and the physical 3D printed headdress. Using 3D scanners and printers enabled the production team to create a unique headdress that was customized to the actress quicker and with more flexibility than traditional methods. Another benefit is that the digital 3D model made for the practical headdress could also be used for the camera work and postproduction CGI.[22]

**Web link**

Watch an interview on the production of the headpiece:

https://youtu.be/6VhU_T463sU

Science fiction television shows and movies are not the only projects using this technology. The musical film *The Greatest Showman*, from Twentieth Century Fox, 3D printed miniature buildings to create their city prop of Brooklyn for a sweeping shot in the film. Production teams decided to use 3D printers to make the miniature city after the studio told them that the cost for the traditional method of creating the miniatures was not in the budget.[23]

With the addition of 3D scanners and printers into the prop production workflow, production studios can also save time and money by building 3D modeling libraries for future projects. This gives them the ability to quickly find and adjust models, fabricate new props, and customize to the production needs.[24]

## Medical

Advances in the use of 3D printers are all over the medical field in unique ways, including medical implants, tools, tissues, organs, hearing aids, casts, corneas, and prosthetics. Every day, more ideas are being developed and breakthroughs are being made using 3D printers to help both humans and animals.[25]

We will focus on the ability of 3D scanners and printers to rapidly prototype items customized to individual patients. For example, in **Figure 2.20** we see the combined use of 3D scanners and 3D printers to make a patient-specific custom cranial implant.

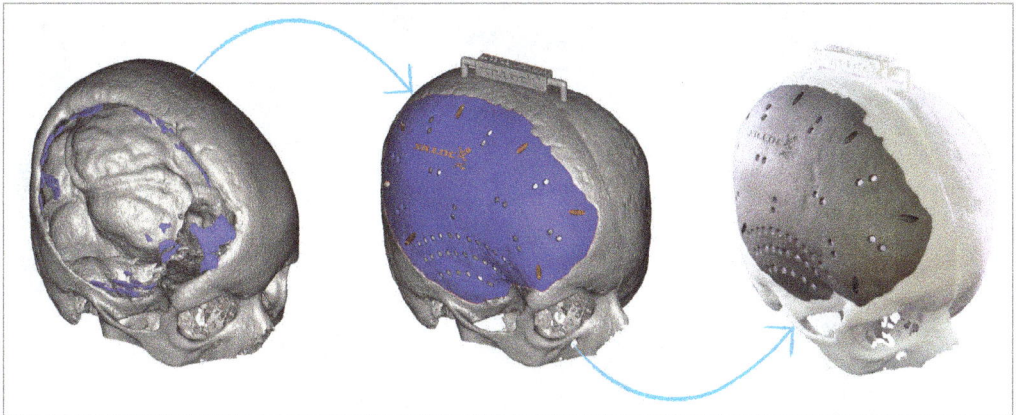

FIGURE 2.20 – From CT-Scan to CAD design to 3D printed Titanium Cranial Implant. Maikel Beerens, Xilloc

## Prosthetics

The design and development of custom-fit prosthetics using 3D printers is growing very quickly. Unlike 3D printing breakthroughs in other aspects of the medical field, innovation in production and customization was driven by DIY makers.

A few massive issues with traditionally made prosthetics is that they are very expensive, lack availability, and take a long time to make. Because of this, children often suffered without them because they grew quickly. They also were not very customizable in both fit and aesthetics.

This started to change with a big rise in affordable 3D printers hitting the market. With the influx of 3D printing enthusiasts, DIY creativity for printing one-off custom objects increased. A few of these DIY makers saw that the 3D printer's ability to rapid prototype could fill a need in custom made prosthetics.[26]

Now, DIY enthusiasts can design and print prosthetics at low cost and without ever meeting the recipient. Makers can obtain common hardware like screws, cable, and fittings and print out the prosthetics based on photographs or 3D scanned models of the recipient. The prints can be scaled as many times as needed to adjust for fit or growing children. The hardware can be reused to rebuild the prosthetic for very little money (see **Figure 2.21**). With the wider availability 3D scanners, the process is becoming more streamlined.[27]

The customization is not only for fit. Makers and recipients have also gotten very creative when fabricating prosthetics. They have built in features, decorations, and even LEGO® mounts. This can create a critical change of thinking for the recipient, from a medical device to a custom extension of themselves.

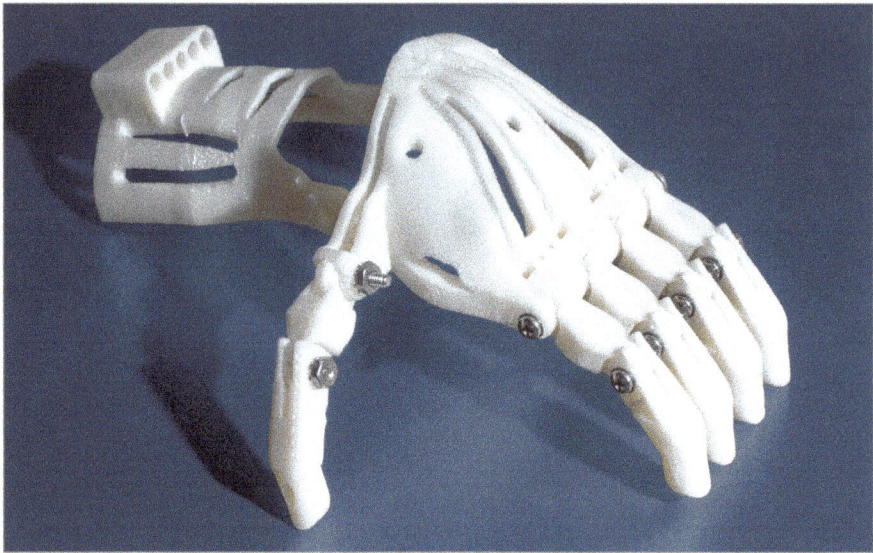

FIGURE 2.21 – This 3D printed prosthetic hand was printed and assembled by FDA researchers in their CDRH laboratory. Similar devices are also available on the internet and are typically used for children born without fingers. The U.S. Food and Drug Administration.

I asked Jen Owen, the founder of enablingthefuture.org, an organization of volunteers that provide free prosthetics that are 3D printed, about her take on the use of 3D printers in the fabrication of upper limb assistive devices.

**INTERVIEW**

**JEN OWEN**

*Founder of enablingthefuture.org and e-NABLE Community Volunteer*

enablingthefuture.org | www.facebook.com/enableorganization
twitter.com/Enablethefuture

**ST: How has the use of 3D printers changed the production and the customization of upper limb assistive devices?**

**JO:** The use of 3D printers has changed the production and customization of upper limb assistive devices by allowing people from all walks of life, different backgrounds and educational levels and areas of expertise to collaborate and share ideas that are creating stronger and more robust low-cost options for people who can't afford a traditionally made prosthetic.

The use of 3D printing has opened up a door for children especially, to help decide how they want their own assistive device to look, (choosing colors and themes etc.) as well as allowed them to be a part of assembling and putting together their own hands and arms. When the children contribute to making the decisions and help to build their own devices, they are more likely to use them and feel like they are part of them.

The 3D printed low-cost basic hands from e-NABLE have allowed children and adults who cannot use traditional prosthetics because they have too much of their hand or palm still, to have an option for using a device that they would not have had otherwise.

Now, anyone that owns a 3D printer or has access to one can print an upper limb assistive device for someone in need.

**Read more about how you can print an upper limb assistive device:**

**www.enablingthefuture.org**

Web link

Digit amputees also benefit from 3D printing. Fully working prosthetic fingers, including the thumb and multiple fingers on a hand, can now be quickly customized and printed for each recipient. This can give added functionality back to amputees that might not have been able to receive a traditional prosthetic.[28]

3D printed prosthetics are not limited to the upper body; the lower body can also benefit from 3D printed prosthetics. 3D printers can print full scale limb prints or in modules to be assembled depending on need.[29]

Animals can also benefit from 3D printed limbs. Domestic and wild animals have received 3D printed prosthetics, including an alligator's tail. An alligator named Mr. Stubbs lost his tail due to illegal animal trafficking. To save the alligator's life, researchers from the CORE Institute in Phoenix and Midwestern University designed and 3D printed multiple tails for Mr. Stubbs to try out and use. Because Mr. Stubbs is still growing, he will need a new tail once he outgrows his old one. Traditional methods would be too expensive, making it ideal for 3D printing.[30]

One more example of the evolution of 3D printed prosthetics is also a look into the future. The video game *Deus Ex: Human Revolution* by Eidos Montréal is set in the future and features a protagonist named Adam Jensen with multiple augmentations, including two prosthetic arms. These augmentations give Denton special abilities. In the game you are able to add modifications to the prosthetics arms to change their abilities.

While the game's augmentation abilities are still in the future, more advanced bionic arms are being 3D printed and customized for recipients. These multigrip arms are advanced with motors, sensors, and electronics to enhance the device. Aesthetic modifications can be added and frequently changed. These bionic arms react to the movement of existing muscles through their sensors to fully move the hand. The company Open Bionics creates cutting-edge bionic arms called Hero Arms. One type of Hero Arm is styled after the *Deus Ex: Human Revolution* prosthetic arms. The future is happening now with bionic arm fabrication with only more advances to come.[31]

**Web link**

**Watch and interview and the Hero Arm in action:**

**https://youtu.be/3nnrstBxomk**

The examples could go on. While the technology and workflow are still growing and adapting, the use of 3D printers and scanners have forever changed how prosthetics can be fabricated while providing DIY enthusiasts a unique way to help others in need.

## Rapid prototyping and manufacturing

As stated throughout this chapter, rapid prototyping and manufacturing are the strengths of 3D printing. 3D scanners and printers are starting to revolutionize the workflow for numerous types of product design and custom manufacturing with both short runs and one-off printing. Using the cheaper and faster method allows the creation of a 3D model, both digital and physical, from a picture or blueprint. This allows for quick adjustments and customization before investments are made in a larger manufacturing processes by companies.

In Chapter 4 we follow entrepreneurs and take a closer look at the rapid prototyping and manufacturing steps. First, we look at one-off prints for special projects. Then we take a look at a case study of the HangTime Hook from start idea to finished commercial product. We also talk with the creators of the Jamber mug and gain insight on prototyping variations and the creative process.

## › 2.2 – What is the potential growth of 3D printer use in industry?

For this book we have only showcased a few areas where 3D scanners and printers have affected our lives and have integrated within different industries. However, the potential growth and existing use is far reaching. Industries and services that you might not associate with 3D printing are probably using 3D technology or have the potential of adding it to the production workflow.

Here are just a few more ways 3D printers are impacting our lives.

- **Agriculture** – fabricating farming tools and polylactic acid (PLA) made from corn starch[32]
- **Armed Forces** – building 3D printed barracks[33]
- **Art** – using 3D scanners and models to make casts for statues
- **Automobiles** – custom made parts printed on the track for Formula 1 cars[34]
- **Environment** – making artificial reefs to help coral and sea life[35]
- **Fashion** – 3D scanning and printing custom shoes[36]
- **Medicinel** – printing replacement corneas, rib cages, and bioprinting organs[37, 38, 39]
- **Sports** – 3D printed titanium horse shoes (see **Figure 2.22**)[40]

FIGURE 2.22 – CSIRO scientists have custom made and 3D printed a set of titanium shoes for one Melbourne race horse in a first for the sport. The horse, dubbed by researchers as "Titanium Prints," had its hooves scanned with a handheld 3D scanner. Using 3D modelling software, the scan was used to design the perfect fitting; lightweight racing shoe and four customized shoes were printed within only a few hours. CSIRO.

The evolution of 3D technology along with the advancements in printable material will continue to drive the growth and integration of 3D printing within industry and or everyday lives. They will continue to speed up production workflows, and lower overall project costs and waste. We have not imagined all the ways 3D printer technology can help yet, but each day brings more discovery.

## SUMMARY

3D printers are no longer just used in large manufacturing industries. They are now also used in small business and consumer markets. 3D printers have made it possible to make objects cheaper, faster, in space, custom made, fix or reimagine objects, and reproduce items from history. They can also make a smaller scale model or make parts to make something bigger. The wide range of uses for a 3D printer in industry is expanding and seems unlimited.

## APPLYING WHAT YOU'VE LEARNED

1. Continue making your own 3D dictionary by adding the definition (in your own words) of five words relating to 3D printing.

2. In what field would you like to make a something with a 3D printer? What would it be, and why would you like to make it?

3. In the different areas mentioned in this chapter, which interest you the most and why?

4. In the year 2050, how do you think the 3D printer will be used?

5. What would you like to reproduce from the past and why?

6. Would you live or work in a building made from a 3D printer? Why or why not?

7. Would you eat food made from a 3D printer? Why or why not?

8. In the medical field, what do you think will be the next big breakthrough? Why?

9. What else can 3D printers do to help the fields of entertainment and sports? Explain your thoughts.

10. What are the needs for 3D printers to advance the fields mentioned in this chapter?

## REFERENCES

[1] – https://www.nasa.gov/exploration/systems/sls/nasa-tests-3-d-printed-rocket-part-to-reduce-future-sls-engine-costs

[2] – https://news.lockheedmartin.com/2018-07-11-Giant-Satellite-Fuel-Tank-Sets-New-Record-for-3-D-Printed-Space-Parts#assets_all

[3] – https://www.nasa.gov/centers/marshall/news/releases/2018/nasa-marshall-advances-3-d-printed-rocket-engine-nozzle-technology.html

[4] – https://www.nasa.gov/centers/marshall/news/nasa-advances-additive-manufacturing-for-rocket-propulsion.html

[5] – https://www.nasa.gov/mission_pages/station/research/experiments/1115.html

[6] – https://www.nasa.gov/content/open-for-business-3-d-printer-creates-first-object-in-space-on-international-space-station

[7] – https://images.nasa.gov/details-KSC-20180316-PH_GEB01_0132.html

[8] – https://ntrs.nasa.gov/search.jsp?R=20180002949

[9] – https://www.aniwaa.com/3d-printing-for-archeology-and-museology/

[10] – https://www.smithsonianmag.com/blogs/national-museum-of-natural-history/2017/11/29/3d-technology-key-preserving-indigenous-cultures/

https://www.aniwaa.com/blog/envisioning-the-future-of-3d-scanning-and-3d-printing-in-museums/

[11] – http://www.bbc.com/news/uk-scotland-north-east-orkney-shetland-37481840

[12] – https://www.aniwaa.com/blog/envisioning-the-future-of-3d-scanning-and-3d-printing-in-museums/

[13] – https://newsdesk.si.edu/releases/smithsonian-releases-high-resolution-3-d-model-apollo-11-command-module-explore-and-print

[14] – http://3dinsider.com/3d-printing-architecture/

[15] – https://3dprinthuset.dk/europes-first-3d-printed-building/

[16] – https://www.naturalmachines.com/faq/

[17] – https://www.sculpteo.com/blog/2017/12/18/dental-3d-printing-how-does-it-impact-the-dental-industry/

[18] – https://www.huffingtonpost.com/entry/3d-printing-technology-transforming-dentistry_us_589cba50e4b0985224db5ea8

[19] – http://www.javelin-tech.com/3d-printer/industry/dental/

[20] – https://www.voxeljet.com/company/news/what-film-and-prop-makers-need-to-know-about-3d-printing/

[21] – https://www.voxeljet.com/industries/foundries/starbug-a-british-tv-icon-is-3d-printed/

[22] – https://3dprint.com/194306/ironhead-studio-3d-printing/

[23] – https://www.fxguide.com/quicktakes/greatest-showman-exclusive-video/

[24] – https://3dprint.com/185871/3d-printed-props/

[25] – https://www.fda.gov/medicaldevices/productsandmedicalprocedures/3dprintingofmedicaldevices/ucm500539.htm

[26] – https://www.theguardian.com/technology/2017/feb/19/3d-printed-prosthetic-limbs-revolution-in-medicine

[27] – http://enablingthefuture.org/upper-limb-prosthetics/

[28] – https://www.tctmagazine.com/3d-printing-news/company-uses-formlabs-3d-printer-to-create-custom-prosthetics/

[29] – https://www.raise3d.com/blogs/case/3d-printed-prosthetic-legs

[30] – http://www.wbur.org/hereandnow/2018/08/17/alligator-3d-printed-tail

[31] – https://openbionics.com/

[32] – https://www.agweb.com/article/the-future-of-3d-printing-on-the-farm-chris-bennett/

[33] – https://www.marines.mil/News/News-Display/Article/1611532/mcsc-teams-with-marines-to-build-worlds-first-continuous-3d-printed-concrete-ba/

[34] – http://www.eurekamagazine.co.uk/design-engineering-features/interviews/how-3d-printing-is-being-used-to-develop-f1-cars-at-the-track/165843/

[35] – https://news.nationalgeographic.com/2017/03/3d-printed-reefs-coral-bleaching-climate/

[36] – https://www.machinedesign.com/3d-printing/creating-custom-shoes-3d-printing

[37] – https://youtu.be/7xoRe2OFNnI

[38] – http://www.wbur.org/npr/440361621/engineers-create-a-titanium-rib-cage-worthy-of-wolverine

[39] – http://www.wbur.org/bostonomix/2017/11/22/3d-bioprinting

[40] – https://thehorse.com/116991/australian-researchers-create-custom-titanium-horseshoes/

# 3D Printers and Education

## OVERVIEW AND LEARNING OBJECTIVES

**In this chapter:**

- 3.1 – How are 3D printers affecting learning in education centers and libraries?
- 3.2 – What is self-guided learning?

There are a number of ways to use 3D printers as learning tools to teach new skills or to learn for your personal knowledge. In this chapter we highlight three education avenues.

- **Education centers** – elementary schools, high schools, higher education, trade schools
- **Libraries** – public and within education centers
- **Self-guided learning** – print publications, online training, conferences, and makerspaces (a space with creative and manufacturing tools)

**IN EXTRAS** *This chapter notes several web URLs. These links are also in the extras and on the DVD in an interactive PDF for quick viewing.*

## › 3.1 – How are 3D printers affecting learning in education centers and libraries?

Education is a broad field including many different areas and grade levels. As with additive fabrication in industry, the use of 3D printers and scanners in education is emerging rapidly into these areas; from building whole curriculums around their use in creative thinking, to having a printer available for individual students, to a tool for expanding their imagination.

In this chapter, we will highlight several ways educators are utilizing 3D printers in education centers and in libraries. We will also take a look at an example of how learning 3D technology can be used in a combination of these areas.

### Education centers

When you first think of 3D technology being used or taught in the classroom, you might think of a computer technology educator teaching about the 3D printer or how to use it. However, the learning for students goes beyond single subjects, learning the software, or how it works.[1]

3D printers are used in many education centers, but the way they are used by the students varies depending on the school's curriculum and resources. Educators see the process of using the 3D printer having numerous benefits for the students, such as developing their hands-on problem-solving, using their creativity to invent, engaging reluctant students to become actively involved, and developing *STEAM* skills. Students can view and hold an object they make. A textbook or lecture will never be able to provide the same experience (see **Figure 3.1**).

FIGURE 3.1 – A group of students interacting with an object they 3D printed. Dolgachov.

## Benefits

Besides teaching the basics about additive 3D technology, educators are using 3D printers as a creative thinking resource in multiple subjects. During the printing process, students and teachers can move from abstract ideas to analysis, planning, problem-solving and review. This engages the student in multiple learning methods, such as visual and kinetic learning. 3D printing a designed object can give the student valuable research and development skills in a hands-on way that a lecture would not be able to replicate. [2]

A few benefits of using 3D printers in the classroom include:

- Creative thinking
- Engaging the student
- Visual and kinetic learning

- Learning through trial-and-error
- Learning how to work through limitations and constraints
- Personalized learning projects
- Job ready skills
- Creating tangible objects from abstract ideas
- Teamwork and collaboration skills

## Wide range of subjects

Numerous subjects can benefit from using 3D printers as a teaching tool. For example, in the field of archaeology, students can now download scanned 3D models of artifacts from museums. Then they can print a scaled replica to examine and hold. For a student learning about history, this brings the past to life.[3] [4]

A few examples of using 3D fabrication technology in a cross-section of subjects are:

- **Architecture/Engineering** – students can design and print prototypes
- **Art** – students can design and print 3D art or adjust their 2D art to print in 3D
- **Biology** – print out anatomical models for study
- **Chemistry** – print and assemble molecular models
- **Computer Science** – learn the related software and hardware
- **Geography** – print topographical maps
- **History** – print artifacts, ruins, and accurate scaled replicas to hold
- **Math** – print geometric shapes and visual representation of math elements
- **Performing Arts** – use 3D printers to print props
- **Writing** – write stories about objects that they 3D printed

## Curriculums

Additive fabrication is also being interwoven into full courses and curriculums. To further explore 3D printers as part of a curriculum and in the classroom, we interviewed Mandi Dimitriadis, the director of learning from Makers Empire, a company specializing in building curriculum and educating teachers and students' 3D printing and Design Thinking.

## MANDI DIMITRIADIS

*Makers Empire, Director of Learning*

https://www.makersempire.com/

https://www.makersempire.com/mandi-dimitriadis-director-of-learning-at-makers-empire-speaker/

**ST: What is Design Thinking and how does it fit in the classroom?**

**MD:** Design Thinking is a non-linear, iterative process that enables us to define and solve problems. It also helps us to reframe problems as opportunities for developing human-centered solutions based on deep insights and empathy for the experiences of others.

In addition, design thinking involves identifying problems, developing empathy, defining problems, generating and visualizing possible solutions, producing and testing prototypes. It is a key aspect of modern design and technology and engineering curricula which focus on the skills and processes students need to develop designed solutions.

Design Thinking is identified as an essential component of the twenty-first century skills students require to thrive and survive into the future, including critical and creative thinking, collaboration, and communication. Teaching students design thinking helps them develop creative confidence and the self-belief that they can solve problems and make the world a better place.

There are many benefits for engaging students in authentic design thinking contexts including:

> Being able to identify problems and reframe them as actionable opportunities
> Understanding the value of collaboration and feedback
> Viewing setbacks and failures as valuable learning moments
> Appreciating the value of hard work and persistence
> Developing self-belief as problem solvers
> Developing empathy
> Developing a growth mind-set
> Developing persistence and resilience
> Developing entrepreneurial and community-minded behaviours
> Having a focus that is both future and solutions-oriented

**ST: How do 3D printers fit into Design Thinking?**

**MD:** 3D printers are extremely useful tools for supporting design thinking processes. They provide an inexpensive and time-effective means for producing prototypes and working models of proposed solutions. 3D printers support the iterative nature of design thinking by enabling students to rapidly prototype ideas, test their prototypes, and make improvements and tweaks to their models. Desk-top 3D printers are relatively portable and affordable, meaning that students can have easy access to tools to produce prototypes and products. In the past, particularly in primary school classrooms, it has been difficult for students to make "real" products, and design projects often result in cardboard models of representations of solutions. With 3D printing, even our youngest students can create authentic, useful products.

**ST: What benefits do you find in using 3D printers in the classroom?**

**MD:** 3D design and printing positions students as creators and inventors, rather than only using technology to consume information. 3D design and printing empower students to bring their ideas to life. If they can imagine something, they can make it. 3D printing enables students to solve real-world problems. Real-world problems don't only include world-scale problems such as poverty and polluted oceans, but also refers to any problem that matters to students, something they care about and is connected to their world.

For example, we are seeing 3D printed products created by young students actually being used to secure leg-straps for a child with cerebral palsy, identify school bags, and hold mobile phones securely on the school ground person's lawn mower.

As described above, 3D design and printing are one of the best tools for engaging students in authentic design thinking contexts. The focus is not just on "making stuff," but on creating solutions that solve real problems or address meaningful challenges. Students use 3D printing to test out and improve on their ideas until they have developed a satisfactory solution outcome.

3D printing can make abstract concepts tangible and can put otherwise inaccessible objects into students' hands, including rare, fragile, or dangerous artifacts.

3D design environments have significant impact on students' spatial awareness and spatial reasoning skills, including the ability to mentally rotate objects and shapes. Spatial awareness is emerging as one of the key predictors of later success in STEM learning and STEM related careers.

**ST: What sorts of skills do teachers and students learn when using a 3D printer in the classroom?**

**MD:** Here is a quick list:

> Spatial skills including mental rotation and visualization
> Manipulation of 3D and 2D shapes
> Technical skills of how to use the 3D printer safely and troubleshoot
> Estimating size and predicting outcomes

> Mathematical skills such as measurement, angles, perspective, and scale
> Creative design skills
> Collaboration and communication skills
> Project and time management skills
> Empathy for the perspectives of others
> Reflection and evaluation
> Design thinking skills

**ST: What kind of reactions do the teachers and students have in learning the new technology?**

**MD:** We find that teachers and students are overwhelmingly positive about embracing 3D technologies. They are excited about having access to 3D printers and the potential outcomes for their students. After the initial excitement, teachers and students report challenging learning experiences and significant learning curves. They feel proud of their persistence and troubleshooting processes. Teachers report significant levels of student engagement when they embed 3D technologies in their teaching and learning. We have met many students who are able to clearly articulate the processes they engaged in with their 3D learning and are proud of their results because of the investment and effort they have made.

**ST: What type of printer(s) do you recommend to use in the classroom?**

**MD:** At Makers Empire we test many different 3D printers to evaluate their suitability for use in primary schools. We focus on safety, reliability, and ease of use. Most of our teachers are primary years classroom teachers, who don't necessarily feel confident using new types of hardware and technology. We advise schools to use 3D printers that have enclosed casings and air filters for student safety. We look for printers that have auto- calibration and touch screens for ease of use. We also recommend that schools use PLA.

**ST: What software do you use?**

**MD:** Makers Empire 3D has developed easy to use 3D modeling software, aimed at 4–14-year olds (see **Figure 3.2**). It is very intuitive and enables young children to begin successfully 3D designing within minutes.

Pretty much anything that is created in Makers Empire 3D is 3D printable and technical CAD skills are not required.

**ST: Do you use or teach 3D scanning?**

MD: This is not something we have included as yet. However, we have seen some interesting examples of schools using 3D scanning technology.

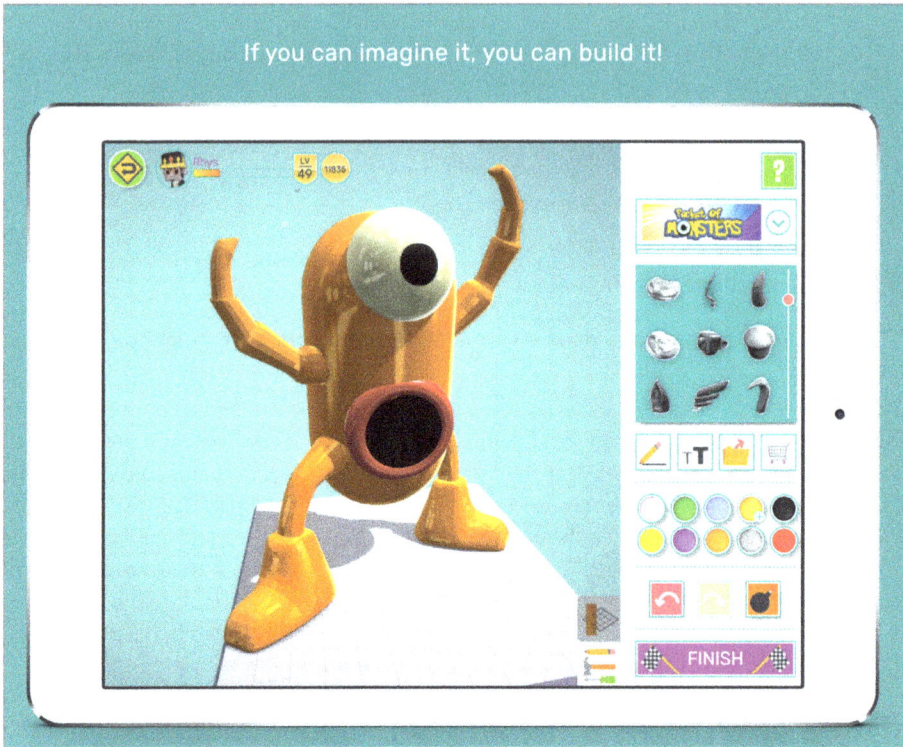

**FIGURE 3.2** – Makers Empire 3D interface. Makers Empire.

**ST: Do you find that teachers and students want to learn more about 3D printing or even make or buy one for personal use?**

**MD:** Nearly every time I visit a school, at least one student asks me how much 3D printers cost and where to buy them from. It seems that 3D printers are appearing increasingly on Christmas wish lists.

Teachers and students always want to know more about 3D printing and often have lots of questions.

I find that people are particularly interested in examples of how 3D printing is being used to help people and do things that would not otherwise be possible. E.g.: prosthetic limbs, medical applications, helping animals, etc.

**ST: What do you see as the future of 3D printers in the classroom?**

**MD:** I believe that 3D printers will increasingly become part of the furniture in everyday classrooms. My vision is that the 3D printer becomes just another tool in students' and teachers' repertoires for creating products, adding specific elements to learning projects and producing prototypes to test and refine ideas.

**ST: Are there any project or stories you would like to share in teaching 3D printing?**

**MD:** Here are some of our favorite case studies:

> https://www.makersempire.com/how-students-used-design-thinking-to-generate-new-town-entrance-sign-ideas-video/
> https://www.makersempire.com/how-3d-printing-united-my-class-school-case-study/
> https://www.makersempire.com/how-birdwood-primary-school-students-solved-a-sticky-problem-with-3d-printing/
> https://www.makersempire.com/how-students-learned-about-design-thinking-by-inventing-gadgets-for-a-superhero/
> https://www.makersempire.com/how-huntfield-school-students-use-3d-printing-to-clean-up-local-pollution/

Here are some of our favorite video stories:

> https://www.youtube.com/embed/qFPu_OJyyTI
> https://www.youtube.com/embed/hN0FD1n-zMU
> https://www.youtube.com/embed/9KHCdDuEW-o
> https://www.youtube.com/embed/Q6S0syM0OFc
> https://www.youtube.com/embed/DKXtfFjNrpo

## Multiple student levels

3D additive technology can be taught in a full range of student learning levels, from elementary schools to higher education.

While learning the inner workings of 3D printers might be too difficult for younger levels, seeing and collaborating during the printing process can spark imagination and creative thinking along with learning basic math and engineering skills. Younger students can use more simplified 3D software to design, while older students can use more advanced software.[5]

Groups and organizations associated with the education centers are using 3D printers to support their activities. For example, a drama club can print props for their next performance, while a robotics club can use the 3D printer to print unique parts for their projects.

Higher education and high schools prepare students for the job force and drive innovation. For example, healthcare simulation devices are effective but costly items. The cost makes it difficult for some education systems to acquire. The solution to this problem was to use 3D printers to print the teaching objects and training tools. This mitigated the cost to just the materials used. The user experience of the newly printed devices needed to be the same for the students as the traditional simulation tools. Rapid prototyping helped researchers to quickly rework the prototypes for functionality and texture so that they could simulate the existing tools. This combined innovation with the development of relevant job skills for the students.[6]

3D printers and other fabrication tools can also bring these levels together. Students and teachers can collaborate on projects to problem-solve or brainstorm. Different grade levels can also teach each other how to use the 3D software and printer.

New education centers are being built around the STEM/STEAM subjects with fabrication labs that include 3D printers.[7] We hear from David Potter about his experience in using 3D printers in STEAM classes.

---

**NAME DAVID POTTER**
STEAM COORDINATOR, DCMO BOCES

**QUOTE**

Students get to hold and see their own products with the use of 3D printers. It is also an opportunity to use the 3D printer to solve problems by designing a part that will benefit a current issue with a product. With time, they will become faster, easily attainable by most and the process of developing or creating with 3D printers will become more of a skill rather than a novelty.

---

**BRIEF REVIEW OF TERMS**

**Science, Technology, Engineering, and Mathematics (STEM)** – refers to the study of a group of fields that include science, technology, engineering, and mathematics.

---

**Web link**

Read more about a newly opened STEM Academy:

http://www.wbur.org/edify/2018/08/23/new-stem-academy

## Integrative technology

Using 3D printers on their own is beneficial, but the student experience and learning can be enhanced when it's combined with other fabrication and immersive technologies, such as Virtual Reality (VR) and Augmented Reality (AR).[8]

For example, a history educator could use a combination of these technologies to not only lecture to the students on the past, but take them on a journey using a recreated world through VR. Students can be fully immersed into the virtual world. They could tour ruins and virtually hold and examine artifacts. Then the class can use 3D models to print those same artifacts to physically hold.

Another example is integrating AR technology to bring physical 2D art to life digitally. An art student can create a 2D image by drawing, painting, photography, or other media, then scan the art to make a digital copy. The artist can then use that digital copy to design a 3D model, animate, and add music or sound effects to enhance the original art. This art field is growing. Art books, shows, and even coloring books (like the Crayola® Color Alive®) are combining AR technology into art. With these digital models, artist can 3D print their creation to bring the digital back into the physical, this time as a 3D representation of the art. [9] [10]

Some workstations, like the HP® Sprout Pro, are all-in-one. These workstations include 2D and 3D scanners, along with other tools such as touch pads and cameras, where students can use to create or modify 3D models to send to the 3D printer. Integrating 3D printers as part of a learning experience with other technologies like VR/AR can expand passive learning into immersive learning.

## Libraries

School and public libraries are including 3D printers and other fabrication technology as part of their collection. One of the fastest growing trends in libraries is to have makerspaces available for their students or the public.[11] These are spaces with fabrication tools and other technologies that help users create innovations (see **Figure 3.3**).

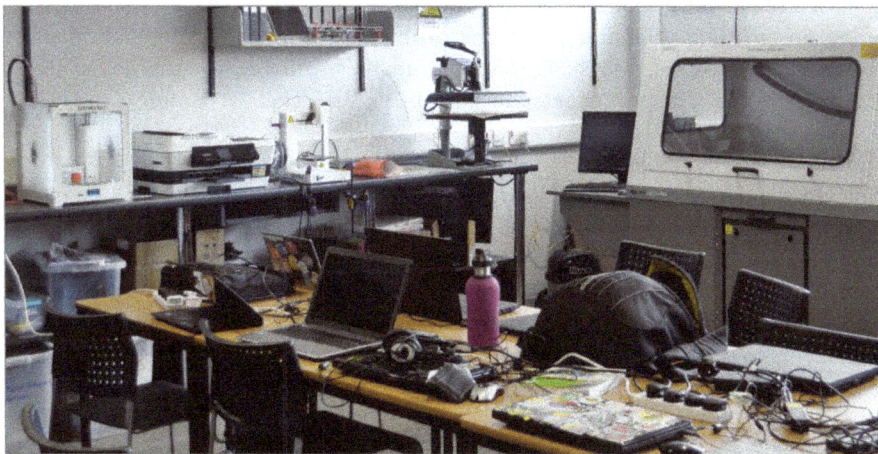

FIGURE 3.3 – The Fab Lab is stuffed with high tech equipment. It is staffed by volunteers who, in return for their work, receive time on the machines. When we visited, the space was also being used as a regional center for the Festival of Code—a group of young people taking part in a national weeklong event, building apps on their laptops. Julia Chandler/Libraries Taskforce.

For school libraries, having 3D technology in a central location allows resources to be pooled and used between levels and classes. This can lower the institution's overall costs for the operation of the 3D printer. Having fabrication technology in a central location helps the school manage the costs of the 3D printer, the materials used, and logistics such as the number of students per printer, teacher training, space available, and time needed for production.

For a closer look into how 3D printers are inspiring middle school and high school students we talked with Dan DeVona, a librarian in a small rural school.

## INTERVIEW

### DAN DEVONA
*MS/HS Librarian, Afton Central School*

**ST: How do 3D printers fit into your course curriculum?**

**DD:** Our library has always been the go-to place for students to create projects that challenge them to discover and to create meaning from what they learn: videos, sculpture, exhibits, floor maps, illustrated chairs! Incorporating 3D design software and a 3D printer added one more tool to do that workshop; with the added bonus of bringing breaking technology to a rural school where many students believe the "cool stuff" happens somewhere else.

**ST: Do you teach using one 3D printer or multiple printers?**

**DD:** We began by piloting a 3D printer on loan from our library consortium. After that introduction, we bought and use a single Printrbot Simple.

I think is important to state that a 3D printer serves the same role as an oven does to a chef. It's really not about the oven or the printer. It's about the whole process of leading up to using it; that's where the problem-solving, the manipulation of ingredients and processes, the skills, and the analysis and revision-discipline are taught and cultivated. The printer gets the glory, but it's the chef we're after.

**ST: What software do you use?**

**DD:** For years we have been using Google SketchUp in our school labs to design and experiment with manipulating 3D objects. It has a low threshold to basic design routines, but also introduces students to the constraints of procedures and the importance of digital disciplines. The fact that it is a free download for home users made it our choice for our 3D design platform. In that our goal in the library is to expose students to the basics of 3D manipulation, prototyping, and printing, it serves our needs. We use Repetier software to configure and print with our Printrbot.

**ST: What benefits do you find in teaching 3D printers to students?**

**DD:** The 3D design and prototype process is a powerful experience for cultivating skills students need to succeed: persistence, attention to detail, self-editing, problem-solving, visualization, analysis, and revision for the best product. Because the actual printing process takes time (and money) it provides a defensible reason for teacher-student "proofing" of designs; just like the "real world." Also, many school projects are about "getting done," more about the ends, rather than about the means. 3D design and printing is very much about the journey; sharing new insights, building on criticism, adapting to obstacles, working at getting better.

**ST: What kind of reactions do the students have in learning the new technology?**

DD: The first reaction to the actual printer is always fascination, a flood of questions, and "can I print something?" Turning the corner from that enthusiasm to a positive first experience using 3D software takes some shepherding, but the reaction to making and "rotating" their first object is clearly an empowering one. As we move through iterations of a student's project, the discussion and rationale of the prototyping process in the real world becomes more fully appreciated and part of their reaction to the technology.

**ST: Do you find that students want to learn more about 3D printing or even make or buy one for personal use?**

**DD:** As in all school-related enterprises, there is the usual bell curve of student expertise and interest. A few have pursued the use of more advanced software and have designed objects of surprising ambition and complexity. I am not aware of anyone purchasing a printer.

**ST: What do you see as the future of 3D printers?**

**DD:** As a tool for learning and for encouraging making over consuming, I would think they would remain a key component of maker spaces. As for real world applications, I assume they will remain an indispensable tool for prototyping ideas if not actually producing customized components.

**ST: Are there any projects you would like to share in teaching 3D printing?**

**DD:** Creating unique chess pieces for our school library! See **Figure 3.4**.

FIGURE 3.4 – 3D printed chess pieces design and printed by High School Students. Dan DeVona.

In public libraries, librarians or their staff can help teach the public who do not have the opportunity or room to buy a 3D printer, support items, or other fabrication tools. The benefits are twofold: the libraries stay current with technology trends, while reaching and attracting a greater audience. The public gains access to technology they might not have the opportunity to otherwise (see **Figure 3.5**).

FIGURE 3.5 – Testing out the 3D printer. Texas State Library and Archives Commission from Austin, TX, United States.

Throughout their history, libraries have filled a gap by providing tools and resources to the public and students. With the addition of 3D fabrication technology into their collections, libraries are evolving with the times and bringing cutting edge technology to their users.

## Combining and Collaboration

The education of additive 3D technology is not isolated to these areas individually. Educators can use a combination of these areas to build upon each other and to help the greater community.

For example, students at Novi High School, Novi Woods Elementary, and the community at the Novi Library in Novi, Michigan, all came together to help children in need by printing and assembling prosthetic limbs for the e-NABLE organization. The e-NABLE organization is a group of volunteers dedicated to use a combined community skill base and resources to 3D print prosthetics for people around the world.[12]

The Novi High School's robotics team was using 3D printers to print and assemble prosthetic limbs for e-NABLE. The team decided to bring more students into the process and mentored students from Novi Woods Elementary in how to print and assemble prosthetic hands.[13] [14]

They then decided to expand the project further to include the public. They met with librarians at the Novi Library to discuss using their 3D printer and space to teach the public about the 3D printer and the process of creating the prosthetic hands.

The combination of education centers and libraries and the collaboration between communities was a great success, with future goals of printing and assembling 200 hands.

**To watch a video showcasing Novi's High School robotics team and the e-NABLE project in action:**

**https://youtu.be/oVqDw_psicU**

**Web link**

We talked with Jen Owen about how learning with 3D printers could not only help the student, but also help others in need around the world.

**INTERVIEW**

## JEN OWEN

*Founder of enablingthefuture.org and e-NABLE Community Volunteer*

enablingthefuture.org | www.facebook.com/enableorganization
twitter.com/Enablethefuture

**ST: Can a new Maker print an upper limb assistive device to learn and to help someone in need at the same time?**

**JO:** e-NABLE has been incorporated into over 2,000 schools worldwide who are using 3D printed hands and arms as a STEM-based service learning project that helps the students learn about the 3D printing technology and design, while also giving them a real-world problem to solve for someone in need of assistance.

When students create a 3D printed hand or arm and send it to a recipient or to e-NABLE to send off to a clinic in an underserved area for their patients to try, the students get to see that their math, design, and art skills that they are learning in the classroom are making a real impact for someone in the real world and not just creating another trinket that demonstrates how the printers and technology work.

For every one child that gets a 3D printed hand or arm from a classroom of students, there are 10–30 students who are being impacted and inspired to find other ways they can use the technology to change lives and help their fellow humankind!

Students at all grade levels are now able to use 3D printers in some schools because of the increase in the focus on STEAM, and the development in curriculum and lesson plans. Their ability is only limited by the resources available at their school. Even though most students may not be using 3D printers when they enter the workforce, schools are preparing students with valuable skills: problem-solving, collaboration with others, perseverance, creativity, and the practical use of technical and mathematical skills.

## › 3.2 – What is self-guided learning?

Another method of learning, rather than the traditional education centers, is self-guided learning. There are many physical and online resources that can be tailored to your learning style, speed, and location. You can also collaborate with other makers to learn tips are help problem solve potential issues. We will cover communities and collaboration areas in Chapter 15 as part of the Knowledge Base section.

### Print publications

There are a plethora of books and magazines dedicated to basic, advanced, and niche markets. While online training and makerspaces are extremely popular, print publications are still a great resource of learning. Books and magazines are physical editions without the need for technology to use even though many have online or digital elements included.

One example of a print publication that includes digital composites is *Make:* by Maker Media, Inc., that includes multiple other benefits, such as online workshops and digital editions with the magazine subscription.

### Online training

One of the fastest growing areas in self learning is online videos and resources. There are countless video channels, websites, blogs, guides, news, and community boards dedicated to all aspects of the equipment and process of 3D printing.

Look at an example of training videos posted on a YouTube channel
Make: by Maker Media:

https://www.youtube.com/channel/UChtY6O8Ahw2cz05PS2GhUbg

Web
link

Some online training sites involve a fee; others are free. You can look for general information or refine you search for more targeted answers to questions. Makers also post projects on community sites for collaboration, troubleshooting, and brainstorming with other makers.

Online training is for educators as well. One example is the Thingiverse Education website that offers over a hundred free lessons for a range of grade levels (see **Figure 3.6**).

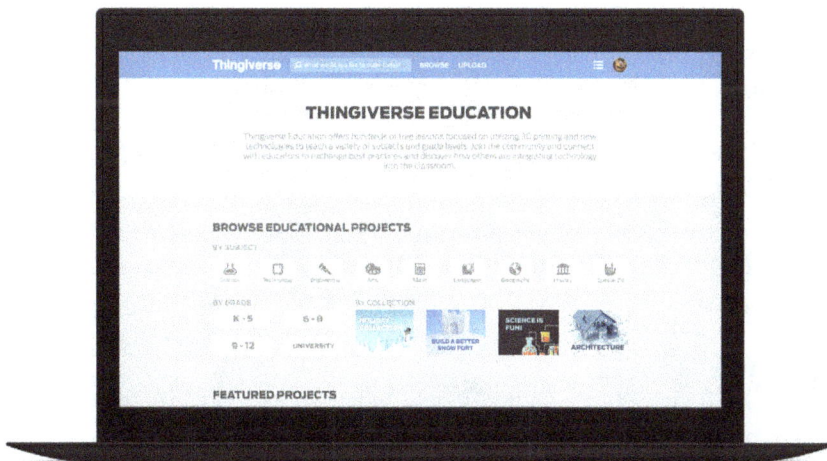

FIGURE 3.6 – Thingiverse Education site. Makerbot.

## Conferences and trade shows

As with other hobbies and industries, 3D printers and other fabrication tools are featured in a number of conferences and trade shows. Makers attend these events to learn more about new advances in technology, popular trends, and collaboration with other makers.

One example is Maker Media, Inc., that sponsors Maker Fairs around the world. These Maker Fairs focus on community creativity and collaborative thinking. They showcase more fabrication tools than just 3D printers and show how using 3D printers in conjunction with other tools can improve or expand the scope of your project.[15]

## Makerspaces and fabrication labs

Buying tools can be expensive or take up a lot of space. A growing trend is to have a public space to house and share a wide range of maker tools. These spaces are a valuable resource for hands-on experience.

Makerspaces can be in commercial spaces or in schools and libraries. Each space might have a different name and will house a different selection of tools to fit economic and space constraints and tailor to the community needs. Some might have instructors, while other have less guided education on the use of the tools. However, despite the differences, the one thing they have in common is that these spaces allow makers to utilize tools and gain practical knowledge they would not otherwise have the opportunity to use or learn.[16] [17] [18]

Examples of these spaces are Fab Lab locations and network. These spaces started out as a MIT outreach program and have grown into a worldwide collaborative network for makers to use tools, learn skills, and work together with other makers (see **Figure 3.7**).

FIGURE 3.7 – Award-winning new library in South Shields, UK. Julia Chandler/Libraries Taskforce.

**Web link**

**Lean more about Fab Labs and the Fab Foundation:**

**https://www.fablabs.io/**

Self-guided training is a valuable resource for any maker to learn new skills at their own pace, skill level, price point, and community group. From the basics to advance problem solving, from the common questions and techniques to the more obscure, a maker can tailor their learning to fit their needs.

## SUMMARY

3D printer usage is different in every school and grade level. It depends on many factors such as: is it in the curriculum or is it an extra activity, are there trained teachers, and how much space and time is available? The cost is also a concern for schools: for example, how many 3D printers can they buy, what kind can they afford, and what is the cost of the materials for the 3D printer? The benefits to students are enormous. Students can now become engaged in authentic thinking and the production of 3D objects. Even if the students don't use 3D printers in the workforce after they leave school, they have learned many valuable skills they need to succeed in life. Collaboration between different grade levels and the community is increasing. Other ways of being educated about 3D printers include, self-guided learning, print publications, online training, conferences, trade shows, markerspaces and fabrication labs.

## APPLYING WHAT YOU'VE LEARNED

1. Continue making your own 3D dictionary by adding the definition (in your own words) of five words relating to 3D printing from this chapter.
2. If you were the teacher, describe a 3D project you would have students make and include the directions plus why you chose the project.
3. If your school were able to expand their 3D printing, what would you like them to do and why?
4. As a student, what do you think will be the hardest part of making a 3D object and why?
5. If your school has a 3D printer, interview the teacher and ask them questions about its use, model, advantages, disadvantages, program, and future use.
6. If you have made a 3D object, what was the best and worst part of making it, and how did you feel when you finished?
7. Describe four skills you can learn from using a 3D printer.
8. Define "design thinking" in your own words.
9. If you could design a project to help the community or school, describe it and how you would make it.

10. If you are going to buy a 3D printer for your home, search on the web to find out which would be best for you (include the advantages, size, cost, and model).

11. If you have made a 3D project, write down all the steps you did and discuss it with the class. If you haven't made a 3D project, find a project using the links in this book and discuss it with the class.

12. What more do you want to learn about 3D printers?

13. List five ways 3D printers have been used in schools.

14. If you had your choice in how to learn about 3D printers, which would you choose—education centers, libraries, or self-learning—and why?

## REFERENCES

[1] – https://edtechmagazine.com/k12/article/2018/02/tcea-2018-how-incorporate-3d-printing-any-lesson-plan

[2] – http://www.eurekamagazine.co.uk/design-engineering-blogs/the-future-of-3d-printing-in-education/158336/

[3] – http://3dprintingsystems.com/why-have-3d-printers-in-the-classroom/

[4] – https://www.blackcountryatelier.com/educational-advantages-of-3d-printing-in-schools/

[5] – https://edtechmagazine.com/k12/article/2015/01/3d-printers-add-new-dimension-classrooms

[6] – https://www.healthleadersmedia.com/nursing/educators-turn-3d-printing-train-nursing-students

[7] – http://www.wbur.org/edify/2018/08/23/new-stem-academy

[8] – https://edtechmagazine.com/k12/article/2018/09/4-ways-k-12-can-maximize-impact-immersive-technology-classroom

[9] – https://youtu.be/mdvVDXpZxqo

[10] – https://youtu.be/eGi7lq5I9K4

[11] – https://lbpost.com/news/education/3d-college-libraries-library-librarians-csulb/

[12] – http://enablingthefuture.org/faqs/media-faq/

[13] – https://www.wxyz.com/news/region/oakland-county/local-students-use-3d-printers-to-create-prosthetic-hands-for-children-around-the-world

[14] – http://www.novilibrary.org/Resources/Access-Technology/Enabling-The-Future.aspx

[15] – https://makerfaire.com/?utm_source=spaces.makerspace.com&utm_campaign=footer

[16] – https://www.makerspaces.com/what-is-a-makerspace/

[17] – https://spaces.makerspace.com/

[18] – https://inventtolearn.com/resources-makerspaces-and-hackerspaces/

by stokkete

# Do-It-Yourself 3D Printing

## OVERVIEW AND LEARNING OBJECTIVES

**In this chapter:**

- 4.1 – When should I use additive manufacturing for my DIY projects?
- 4.2 – How are entrepreneurs using 3D printers?
- 4.3 - What can hobbyists use 3D printers for?

In this chapter, we will take a closer look into the use of additive manufacturing by do-it-yourself entrepreneurs and hobbyists.

- **Entrepreneurs** – Individuals and small businesses that use 3D printers and services to prototype or make commercial products
- **Hobbyists** – Individuals or groups that use 3D printers for home use, to learn new skills, or for fun

We will experience firsthand accounts from DIY consumers and makers in each section to give some insight to how they used 3D printers to solve problems, save time and money, and enhance creativity.

**IN EXTRAS**

*This chapter notes several web URLs. These links are also in the extras and on the DVD in an interactive PDF for quick viewing.*

## › 4.1 – When should I use additive manufacturing for my DIY projects?

For entrepreneurs and hobbyists, there are general questions that should be answered before using additive manufacturing for creating a new object or product. 3D printers are tools and are most efficient when used to their strengths such as speed, flexibility, and overall cost.

Some questions you should consider:
- What is my budget?
- Is this project going to be a proof-of-concept prototype?
- How many prints will be made?
- Are there premade 3D models that you can use or modify?
- What is the 3D print going to be used for?
- Does it require a specific level of strength?
- What type of material can be used?
- What other tools are available?
- What is the timeframe of the project?

Do the answers to these questions point to using a 3D printer as a tool for your project? If so, there are still a few options to consider to best fit with the needs of your project, such as using a print service or investing in your own printer. The following firsthand accounts showcase the decision making, problem- solving, and thought processes done by select entrepreneurs and hobbyists as they used 3D printers for their projects.

## › 4.2 – How are entrepreneurs using 3D printers?

One of the main strengths of additive manufacturing is rapid prototyping. As in large industrial applications, this benefit also assists small businesses and individual entrepreneurs. Concepts can become tangible and projects can be modified quickly and cheaply while maintaining quality. In addition, with the overall cost of using 3D printers and services dropping, entrepreneurs can use them more effectively for small and larger production runs.

- **Rapid prototypes** – one-offs and proof-of-concept prints, used as stepping stones to full production
- **Small lot production** – smaller print runs for testing, initial sales, or to raise capital
- **Business production** – generally larger runs with more resources for usability testing, initial sales, to raise capital or end product for consumers

## Rapid Prototypes

As covered in the previous chapters, using additive manufacturing for rapid prototyping concepts or one-off products can be a powerful tool for both the consumer and the entrepreneur.

In this section, we learn from firsthand accounts of rapid prototyping and from an entrepreneur and customer for a one-off project. We get a glimpse of how 3D printers are interwoven in our daily lives.

### *Firsthand Account:* Rapid Prototyping

### DIY Entrepreneur: Tony Hu

If you are working in creating a proof-of-concept or prototype variations of your project, 3D printers are ideal to use. Our first account is an interview with maker Tony Hu who works for both MIT and Brainy Yak Labs and uses 3D printers during the production phase of process.

**INTERVIEW**

### TONY HU

*Academic Director, Integrated Design & Management Program, MIT; Chief Yak, Brainy Yak Labs*

www.brainyyaklabs.com

**ST: How do 3D printers fit into your work/class projects?**

TH: I use them for prototyping.

**ST: What benefits do you find in using 3D printers as part of your work/class projects?**

**TH:** It allows me to make prototypes more quickly at lower cost. I can more easily fit in more design iterations and get more feedback from users.

**ST: How does the use of 3D printers affect your production timeline?**

**TH:** It decreases the timeline somewhat, but it's hard to quantify. In some cases, 3D printing might allow for more prototyping and iteration within a similar overall timeline.

**ST: Was there a cost savings versus fabricating the project in a different method?**

**TH:** There is often a cost savings and a time savings over other methods. In the past, I might have a supplier in China machine a prototype which might take a week to fabricate and 2–7 days to ship. Depending on the service, 3D printing might take 1–7 days. I have 3D printed using 3D Hubs locally, Shapeways, and suppliers in China.

**ST: How did you decide to use a 3D printer for projects?**

**TH:** Faster, better, cheaper.

**ST: How does the 3D printer work in conjunction with other fabrication tools?**

**TH:** We use a variety of tools depending on the need: CNC mill, laser cutter, hand sculpt, casting, etc.

**ST: What iterations do you go through for a typical project? What did you change during the process?**

**TH:** Many iterations: sketches, renderings, rough protos, more refined protos, pre-production protos. Some changes can be very significant.

**ST: What type of printer(s) do you use? Did you build your own?**

**TH:** We use various printers from service bureaus and makerspaces.

**ST: Do you start with a physical model, a drawing, or a CAD design?**

**TH:** We typically start with drawings.

In school, we teach students to start with quick sketches or hand-made sketch models (quick prototypes). The temptation is to start with CAD and maybe 3D printing, but this is much more time intensive. Sinking time into a design makes you feel invested in it and less likely to consider alternatives. You might be able to generate 10 or 20 ideas on paper in the time it takes to create one CAD model.

**ST: What software and material do you use?**

**TH:** Illustrator, SolidWorks, Creo, many materials.

**ST: What are some unexpected things you learned during the building process?**

**TH:** We always learn unexpected things when testing with users. They often use our prototypes in unexpected ways. That's why it's critical to test with users throughout the development process.

**ST: What do you see as the future of 3D printers?**

**TH:** I'm excited about advances in printing with new materials and the increased use of 3D printed parts in production products.

## Firsthand Account: One-off Headphone Horns

For this firsthand account, we hear from both sides of the project. Consumer Jen Crovo and entrepreneur Jonathan Torta give their thoughts on creating a one-off project to solve a problem.

### DIY Consumer: Jen Crovo

I had sculpted a set of horns as decorative attachments for my headphones (see **Figure 4.1**). The challenge I discovered was in figuring out a way to attach the horns to the headphones securely yet not permanently.

I still wanted a way to be able to remove the horns in case the headphones broke or I wanted to update the decorative attachments. My initial idea was crafting hooks that would grab the curved edge of the headphone bell; while this worked, it was not ideal and due the uneven edge along the horn, it was not easy to form the hooks properly and the horn would slide around the bell. The hooks were also not secure and made from a heavier material. The most important requirement of the horns was that they needed to be lightweight.

I discussed the issue with Jon, and with his help, we were able to create a clip. He 3D scanned the circular base of the horn (the area that would attach to the headphone bell) and the outer edge of the headphone bell.

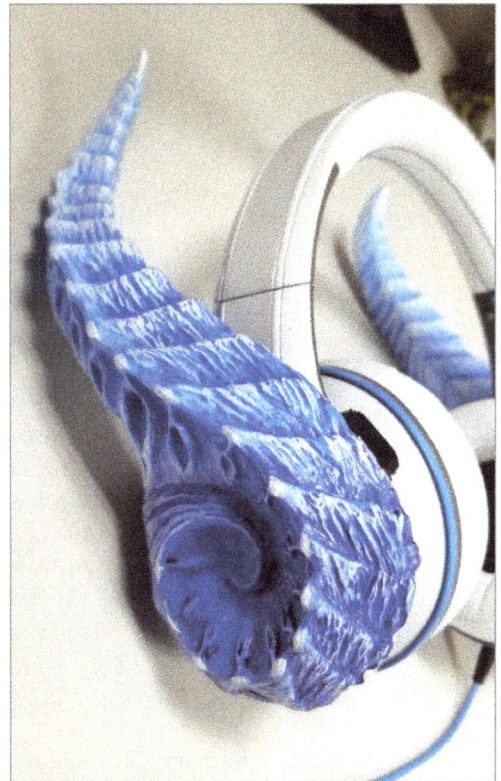

FIGURE 4.1 – The sculpted custom horns. Jonathan Torta.

With the two opposing sides in the 3D space, he was able to create a three-pronged clip that both matched the inner side of the horn base and the outer curve of the bell (see **Figure 4.2**).

The clip was then 3D printed and glued to the base of the horn. The clip securely holds the horn to the headphone with no wiggle room due to its exact contour recorded from the 3D scan. The clips work exactly as I imagined! (See **Figure 4.3**.)

FIGURE 4.2 – 3D model of the headphones, attachment, and horns. Jonathan Torta.

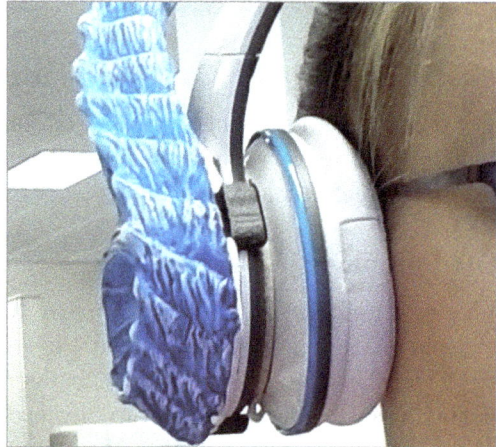

FIGURE 4.3 – Finished product with the 3D printed attachment holding the horn in place. Jonathan Torta.

### DIY Entrepreneur: Jonathan Torta

I was asked to problem solve a DIY project including headphones and two attachments horns.

Jen had other attachments previously and those were held together rather precariously with a bit of *Worbla* (a thermoplastic) and hooks. The requirements were: not to harm the headphones and the horns had to be removable.

Jen's new horns were sculpted and cast in foam. Being hand-sculpted and cast made each horn unique. This also created a problem. Her previous manner of attachment no longer would work and she had to find a new way of attaching the horns. This is where I came in. I offered to make a 3D printed mounting plate that would snap on to her headphones and the horns could be glued to.

This was the start of my interesting project build. I first thought of just drafting the plate in Fusion 360 and having her fit the horn with attachment posts and filler as needed. This was less than ideal as it required the horns to be farther out from the headphones, to have small holes drilled into them, and potentially alignment problems. An additional problem was not having direct access to the headphones for very long as she needed to use them.

So, my new course of action was to 3D scan the headphones and the horns for reference. Normally, I would just take photographs; however, the horns were tricky (see **Figure 4.4**).

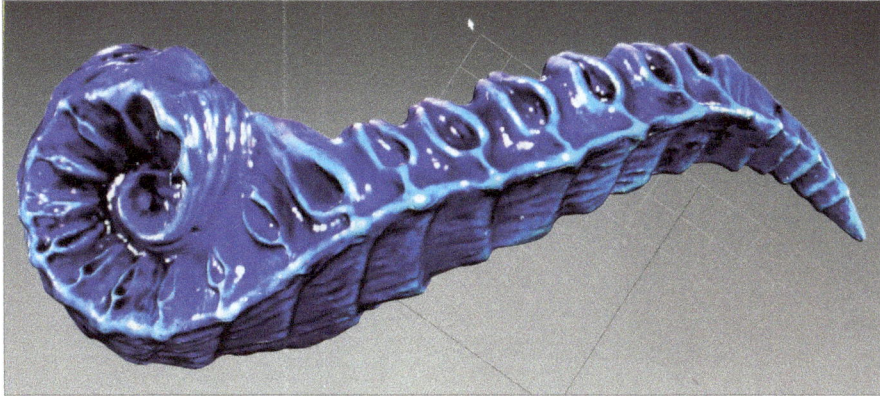

FIGURE 4.4 – 3D scan of the custom horn. Jonathan Torta.

After scanning, I had to reduce the number of faces of the resulting 3D meshes by great degree to be able to be loaded into a CAD program. I did this by chopping out all of the mesh that I did not need. For example, I just kept the attachment area (see **Figure 4.5**).

I then imported the *mesh* and used that as a reference and for dimensioning to make the mounting plate. This mounting plate had a flat back surface which followed the curve of

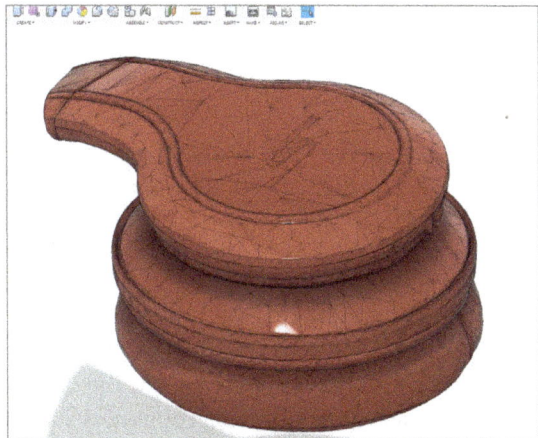

FIGURE 4.5 – 3D model of the headphone attachment area. Jonathan Torta.

### BRIEF REVIEW OF TERMS

**Mesh** – A representation of a given shape or form, consisting of an arrangement of a finite set of geometric components

**Worbla** – A brand of thermoplastics designed for sculpting the headphones and had three clips that reached around and held the plate on (see **Figure 4.6**). This surface is where I needed to mount the horns. Both of them were different and irregular. It was very hard to draw anything that would fit appropriately.

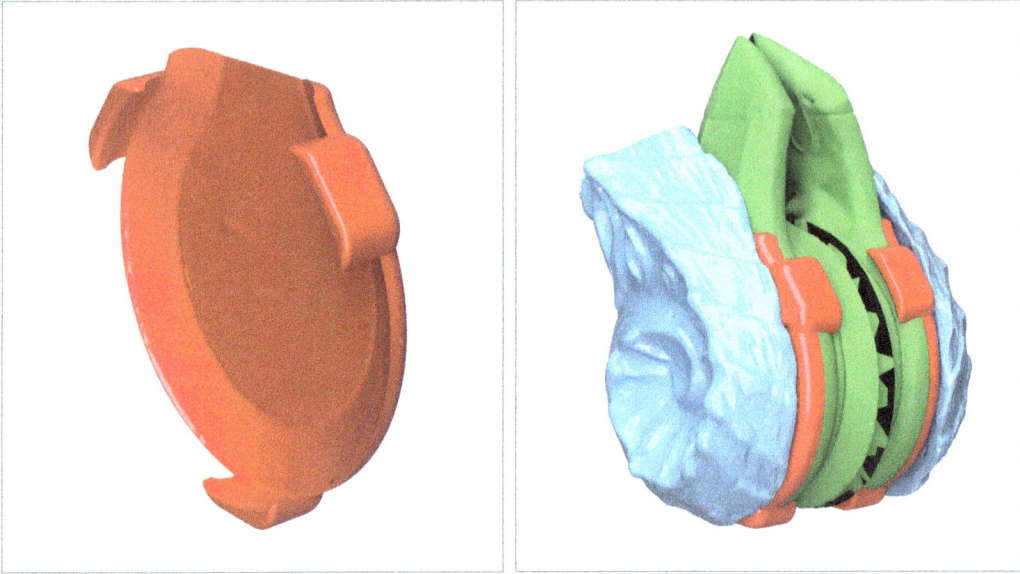

FIGURE 4.6 – A. Mounting plate   B. Mounting plates attaching the horns and headphones. Jonathan Torta.

I then thickened the back of the mounting plate. Afterwards, I used the scanned mesh of the horn and did a Boolean cut into the mounting plate. This effectively cut into the mounting plate a copy of the base of the horn. I'll have a mounting plate custom designed for that specific headset and for the unique right and left horn.

Now to print, this was potentially going to be an issue because there was no good orientation to print the mounting plate where one side had the clips and the other side was very irregular. But since the horn was to be attached to the regular side and additional roughness would be a beneficial to the gluing process, I printed with the irregular side down with supports. This created a slightly rough side that would be easy to glue to the horn. I was also sure to print with three to four *shells* for additional strength. See **Figure 4.7**.

Additionally, I could have printed this in *PETG*, but it seemed to have enough strength and function adequately in *PLA*. When printed, Jen now had two mounting plates for her horns. All she needed to do was to glue the horns on and paint the clips to whatever color she wanted.

FIGURE 4.7 – Prototypes of the 3D printed mounting plates. Jonathan Torta.

---

**BRIEF REVIEW OF TERMS**

**Polylactic Acid (PLA)** – Type of common FDM Thermoplastic filament

**Polyethylene Terephthalate Glycol (PETG)** – Type of FDM Thermoplastic filament

**Shell/Wall** – The outer wall of a model. Shell thickness refers to the number of layers that the outer wall will have between the outside and the infill (think of this as the skin of the object). The higher the setting is for shell thickness, the thicker the outer walls of your object will be

---

## Small Lot Production

While one-offs and prototyping are usually lower print runs, 3D printers can also be used effectively for small lot production. The production process might still involve proof-of-concept and prototype variation prints; small lot production will take the project to the next step in manufacturing.

### *Firsthand Account:* HangTime Hook

### DIY Entrepreneur: Eric Johnson

Here we talk with Eric Johnson and his process to design and produce the HangTime Hook. 3D printers were used for rapid prototyping and small lot production to test usability and raise capital.

## ERIC JOHNSON

*CEO Ridgeline Media Systems LLC* | https://ridgelinemediasystems.com/

**ST: Can you tell us a little about the HangTime Hook?**

**EJ:** The HangTime Hook functions as an organizer and supports more than just media devices! It can support action cameras, aim flashlights and fans, even hold closed containers, drawstring bags, accessories, manage charging cords and headphone cables! It was very handy, universal, and functional, yet simple and light weight! (See Figure 4.8.)

FIGURE 4.8 – The finished product is ready for the consumer. Eric Johnson.

With a 3D printer my functional prototype was distributed as a rough beta model (see Figure 4.9). This not only demonstrated the strengths and weaknesses of the design to improve for real world use, it was essentially grass roots marketing.

I distributed the product at a cost that was affordable to the consumer, was close to the market value of production units, and enough were sold to finance an attorney for filing a Non-Provisional Utility Patent, and injection mold tooling!

Of the initial 8,000 units produced, 6,600 were pre-ordered. In the two weeks since they have been on retailers' websites, I have less than 1,000 remaining and now distribute to six retailers in the US, Canada, Germany, and Australia. I'm writing invoices for as much as $10,000 to $20,000 for parts orders! Enough to pay off the manufacturing costs and commission another product run. I would have to say, to go from proof of concept, to field testing, to marketing, to injection molding production in one year without investors or conventional crowd funding (Kickstarter or Indiegogo) it would not have been possible without 3D printing.

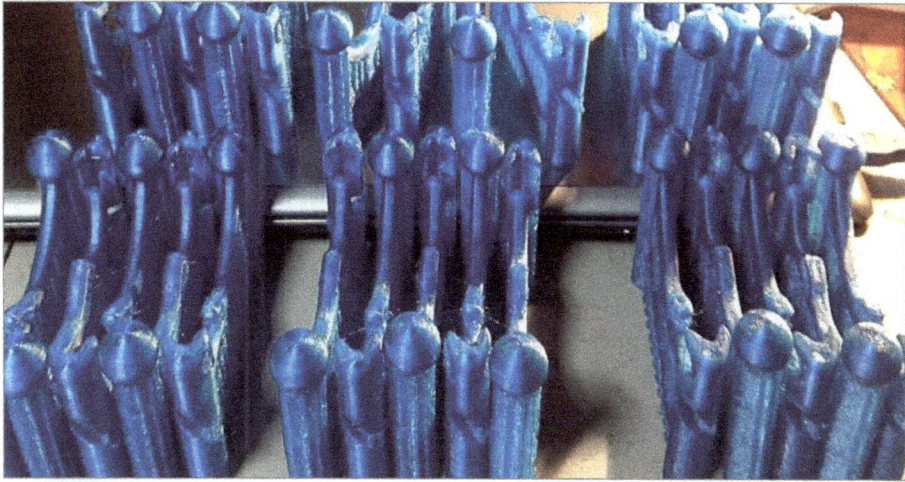

FIGURE 4.9 – 3D printed HangTime Hooks. Eric Johnson.

**ST: How did you decide to use a 3D printer for your project?**

**EJ:** The expense, expedience, and affordable off-the-shelf product coupled with fast turnaround and relative ease of use made it the obvious choice.

**ST: What benefits do you find in using 3D printers as part of your project(s)?**

**EJ:** I reached out to someone for 3D prototyping and the turnaround was fast, even faster when I got my own! I can alter or modify the design and test it in reality in only a few hours! (See Figure 4.10.)

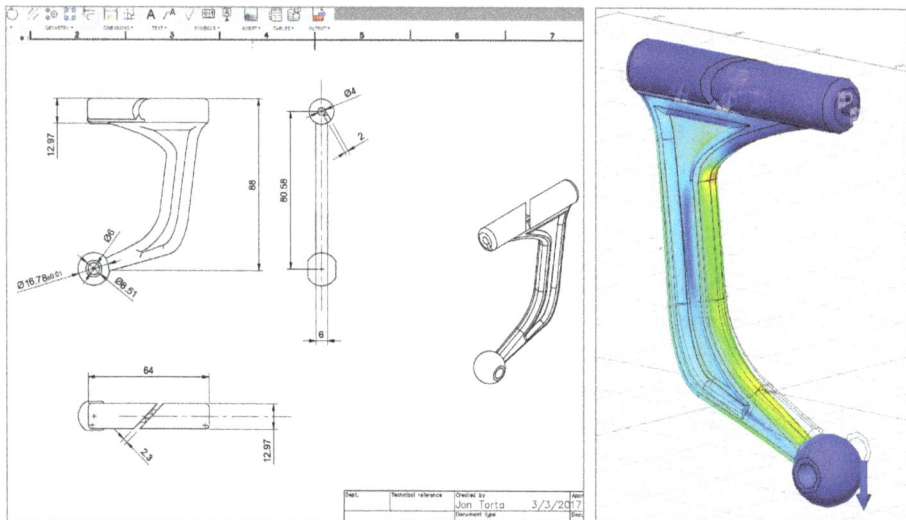

FIGURE 4.10 – The CAD drawings and 3D model for a prototype of the HangTime Hook. Jonathan Torta.

Within hours of modifying or designing a concept, it may be brought into existence and tested! (See Figure 4.11.)

The prototypes I made allowed me to test them in the real world, and develop the concepts rapidly. Eventually I was able to sell beta versions of a product to raise capital for production.

**ST: Did you start with a physical model, a drawing, or a CAD design?**

**EJ:** The proof of concept and more dynamic prototypes were initially hand-shaped physical models that CAD designs were based on. Once I had CAD designs it was as simple as uploading to a 3D printer, and so was limited production!

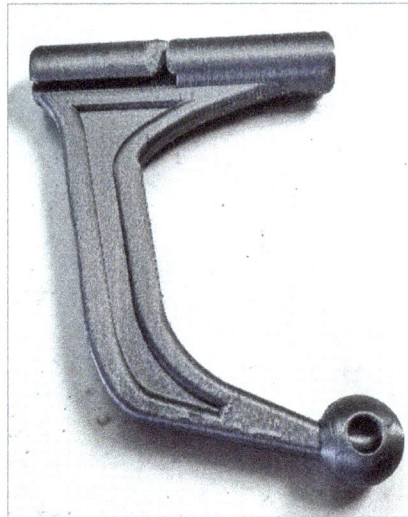

FIGURE 4.11 – The printed prototype of the HangTime Hook. Jonathan Torta.

**ST: What type of printer and software do/did you use?**

**EJ:** XYZ Printing Davinci Pro with stock software for the XYZ Printing Davinci Pro.

I did not go the building route. I purchased an enclosed XYZ Printing Davinci Pro that could receive third party filament and it was virtually plug and play.

I was surprised that though I did not build my 3D printer I had to modify it, use standardized fittings for the Bowden tube and that facilitated using PETG in addition to PLA and ABS. I also got more consistent prints with all materials afterwards.

**ST: How did the use of 3D printers affect your production timeline?**

**EJ:** I went from proof of concept to full production within one year, which allowed the formation of a startup company and sales around the world through cottage industry (niche) retailers.

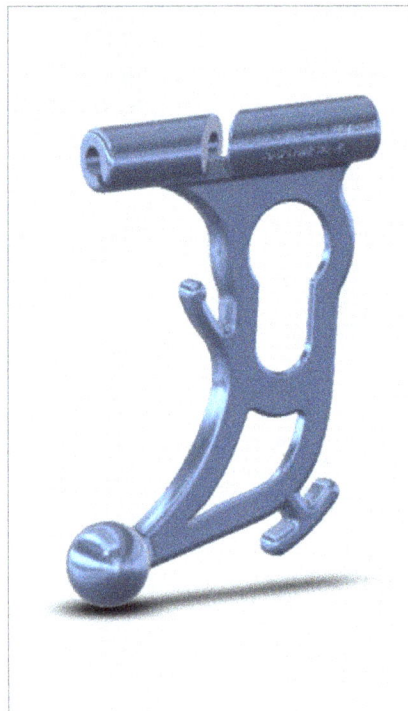

FIGURE 4.12 – Redesigned prototype of the HangTime Hook. Jonathan Torta.

**ST: What iterations did you go through?**

**EJ:** Maybe three – proof of concept, redesign for more dynamic function, and then incorporating adaptability into the final model (see Figure 4.12).

Once I understood how the material behaved and dialed the settings, generally it was predictable enough to have successful prints within the first print. Anything beyond that was simply a matter of maintenance such as cleaning the nozzle or the surface of the bed for adhesion.

**ST: What did you change during the process?**

**EJ:** The most notable change was material. I switched from ABS to PETG for a more resilient prototype that was somewhat flexible while being far less brittle, and featured far better layer adhesion and durability.

**ST: Was there an overall cost saving?**

**EJ:** Absolutely, without a middleman or skilled labor I was able to create parts myself, in my kitchen, at any moment an idea struck. This is as inexpensive as the convenience is priceless.

As opposed to hiring a machinist, sculpting ... absolutely. The closest to rapid production would have been poly-morph plastics that could be heated and molded by hand, but that required finish work and would not be consistent from one prototype to another without a physical mold. Casting was also an option, but a forge would be a bad idea on my kitchen table, and the preparation, fumes, and hazards would be many times more problematic.

**ST: What do you see as the future of 3D printers in your business?**

EJ: Rapid prototyping; 3D printing has proven itself to those ends fantastically.

I should also add that unlike injection molding I was also able to run many different colors in a short time with less prep and expense. This allowed me to determine the most popular colors effectively by polling paying individuals and demand. I could even make special runs of unique colors for season or as limited editions increasing value with rarity.

The speed and volume of the printers as they exist now is under par, I do expect this will change. Most importantly the layer adhesion issues as compared to the density of pressurized injection molded plastics and the consistency of the prints due to ambient temperatures and humidity during the print are things that would need to be improved for production of reliable parts. A secondary finish process was also necessary due to the supports for printing whereas injection molding was finished in one shot.

## Business Production

Like rapid prototyping and small lot production, small businesses use additive manufacturing as a tool for the initial proof-of-concept and prototype testing. The manufacturing process is often expanded due to increased resources or scope of projects.

With larger businesses, customization becomes more of an option. Automotive and aerospace manufactures are examples of companies developing their own additive manufacturing processes and tools to fit their production needs.

### *Firsthand Account:* Jamber Mug

**DIY Entrepreneur: Allen & Diana Arseneau**

For a closer look into the use of 3D printers in developing new products, we talk with Allen and Diana Arseneau, co-founders of Jamber, Inc. They develop innovative solutions to improve upon consumer products and enhance usability for the consumer. In the following interview they talk about their first product, the Jamber mug.

---

**INTERVIEW**

### ALLEN & DIANA ARSENEAU

*Husband-and-wife co-founders of Jamber, Inc.*

www.jamber.com

**ST: How do 3D printers fit into your business?**

A&DA: Jamber creates innovative consumer products through a unique human-centric data-focused design process. This would be impossible, or prohibitively difficult, without the use of 3D printing because our design process is based on mathematically modeling physical interactions with consumer products. The curves of our products are often defined by high-order polynomial equations that simply cannot be modeled by hand. We need 3D printing to transform our math equations into physical products with exacting specifications.

Jamber enhances the lives of everyday people by innovating consumer products that have been overlooked, using a science-driven design approach, called human-centric data design. We believe that everyday products should enhance life without sacrificing beauty or function. Our first products are bio-engineered coffee mugs that could only be designed with the help of 3D printing (see Figure 4.13).

Human-centric data design requires using actual data, based on how the human body interacts with physical products, to define the shapes, curves, and thicknesses of the products, instead of relying on just intuition, preference, or style. To accomplish our

---

FIGURE 4.13 – Jamber Coffee Mug, Confetti 4-pack. Photo image courtesy of Jamber, Inc.

goal, we studied the biomechanics of how humans interact with products such as a coffee mug, and created a family of high-order polynomial equations that dictate the Jamber mug design. 3D printers were vital to transform these equations into real physical products.

Normally, ceramic molds are sculpted by hand. This is the way ceramics have been made for millennia. Bioengineering a ceramic coffee mug required a more exact process. We used 3D printing to create rapid plastic prototypes that we could quickly pick up and test for general feel and look. We also used 3D printing to create Jamber models that were then used to create molds that are used to create the ceramic mugs. This 3D print-to-mold process allowed us to fairly quickly and cheaply get ceramic samples into the hands of our customers for real user testing. This same process was then used to create the 3D printed models that were used to ultimately create the production-ready molds for the actual production process in the ceramic factory.

3D printing has allowed us to bio-engineer a better coffee mug that is literally enhancing the lives of people around the world – from professional athletes to office workers to those suffering from arthritis, Parkinson's, or carpal tunnel. Customers from all over the world contact us almost every single day, and share stories about how their Jamber coffee mugs are enhancing their lives. It makes the 2.5 years of R&D that went into the Jamber mug worth it.

**ST: What benefits do you find in using 3D printers as part of your business process?**

**A&DA:** 3D printing is an incredibly fast, cost-effective, and secure way to design unique products. We are able to take any design that we come up with, and turn the exact design into a physical product that can be handled and tested, generally in as little as about 12 hours, at a cost of about $10 in material and electricity, without having to share our intellectual property (i.e., our designs) with anyone outside of our company. In contrast, having prototypes made externally generally costs hundreds of dollars per sample set, takes several business days to weeks (depending on scheduling and shipping), and exposes our IP to potential competitors (see Figure 4.14).

FIGURE 4.14 – A. Playing with scale  B. A Jamber Mug for the Holiday Elf. Photo images courtesy of Jamber, Inc.

This means that we can look at prototypes quickly, make design changes that same day, and then make more 3D samples. This accelerates the design process dramatically (see Figure 4.15).

**ST: What type of printer, software, and material do/did you use?**

**A&DA:** We use an Ultimaker 2. We use Simplify 3D, and we use PLA 2.85mm diameter.

**ST: How did the use of 3D printers affect your production timeline?**

**A&DA:** Using a 3D printer in our design process has dramatically accelerated our production timeline. Instead of sending designs out to a mold maker, which costs

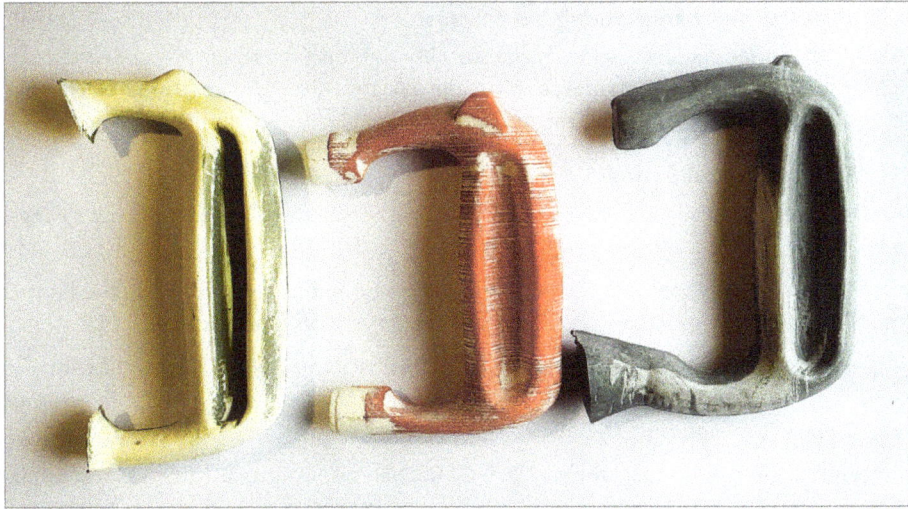

FIGURE 4.15 – Printing for production   A. far left: outsourced 3D printed handle   B. middle: 3D printed handle by Jamber using PLA and the Ultimaker 2   C. far right: epoxy handle from 3D print for making master molds. Photo image courtesy of Jamber, Inc.

hundreds of dollars per prototype set, and often takes days to weeks to complete, we were able to 3D print our own prototypes for about $10 (in material and electricity) and within 12–24 hours, depending on the size and quality of the print. We believe it would have been prohibitively difficult to design the Jamber mug without 3D printers.

We also used 3D printers to speed up the production mold-making process. Sending our factory a 3D printed prototype allowed them to avoid having to handcraft a model, which would take them five business days. They would then have to send that sample to us and wait for our feedback before starting the next phase of production.

### ST: What iterations did you go through?

**A&DA:** We went through nearly 100 design iterations over a 2.5-year period. During this period, we oscillated between research and development, user testing, data analysis, and product design phases. User testing was definitely the most important, and most difficult, aspect of our design process. We created a feedback loop that allowed us to hear directly from customers and decision makers, allowing for rapid design modifications using our 3D printer.

Jamber mugs spent about 1.5 years in various user-testing studies, throughout the 2.5-year design process. These studies included people of all abilities – young and healthy, as well as the impaired. This ensured that the Jamber mug got thoroughly tested by a very high-need population with Parkinson's disease, Alzheimer's disease, osteoarthritis, rheumatoid arthritis, carpal tunnel, and many other limitations. And Jamber mugs were tested by Millennials so that we received wide-reaching feedback. The results were wonderful – our simple coffee mug actually improved people's lives.

**ST: What did you change during the process?**

A&DA:We added features, such as the stability nub and flared lip of the 8oz teacup, and tested different sizes, shapes, and handle designs. After 100 iterations, the Jamber mugs went through some pretty significant changes!

**ST: Was there a cost savings?**

**A&DA:** There certainly was cost savings, but the real benefits are in the time savings, and in our IP protection.

If we were to assume we printed 100 prints during the design process for the Jamber mug, our cost savings from using our own 3D printer would be about $36,000.

We estimate a print costs us about $37. This consists of about $7 for material and electricity, and about $30 for the cost of the 3D printer itself (our printer cost us ~$3,000). We are ignoring the cost of our time.

In contrast, it would cost about $300–$500 per print/sample (an outsourced 3D print or an outsourced ceramic sample). This means that we would have spent about $40,000 for those same 100 prototypes. Instead, we only spent about $3,700, for a savings of about $36,000.

**ST: How did you decide to use a 3D printer for the project?**

A&DA: We are creating shapes that simply cannot be replicated by hand. We had to use 3D printing technologies. We didn't have any experience with 3D printing before starting, but we knew we had to learn quickly.

**ST: Did you start with a physical model, a drawing, or a CAD design?**

**A&DA:** We started with Play-Doh and sculpting clay. We moved to 2D sketches, and then finally moved to 3D files using Solidworks.

**ST: What do you see as the future of 3D printers in your business?**

**A&DA:** 3D printing will continue to be a staple for our product development, prototyping, and manufacturing process. We only wish we could find a good excuse to purchase a chocolate 3D printer. Chocolate Jamber mugs would make amazing holiday gifts!

**ST: Are there any project or stories you would like to share in utilizing 3D printing?**

**A&DA:** The day we bought our 3D printer was exciting. We had thoroughly evaluated resolution, nozzle speed, filament diameter, and other specifications we felt would best suit our needs. The day our printer arrived was even more exciting. But then we discovered the biggest shocker about our 3D printer – the archaic way that it is calibrated. When we called in for troubleshooting and help calibrating, our 3D print guy told us we need to calibrate by hand, using a piece of printing paper, and to manually adjust the print plate so that the nozzle just barely touches the paper on the print plate, but does not scrape it. That's how we calibrate?! Looking back, we shouldn't have been so shocked. 3D printing is still in its early years, even though it's been around for decades. But, at the time, we were surprised that a machine that we had purchased for precision, had to be calibrated by hand with imprecise criteria, and that failure to calibrate properly was one of the biggest reasons for print failures.

Oh! And the failures. We had so many print failures. We have kept our most beautiful failures at our office to remind ourselves of the beauty and learning that comes from failure (see Figure 4.16).

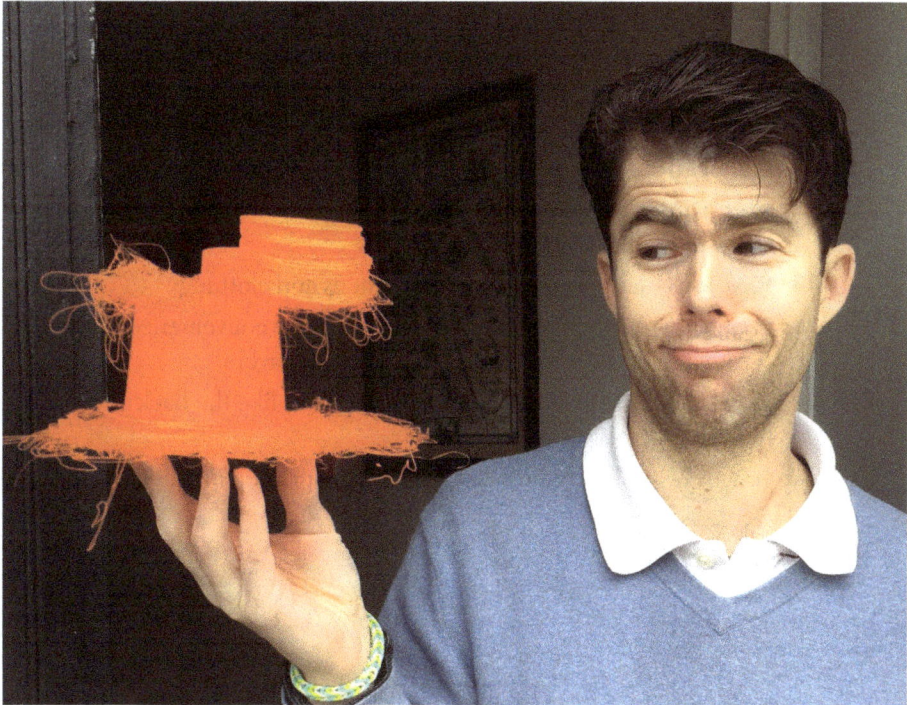

FIGURE 4.16 – Frustration after finding a 26-hour printing "Beautiful" failure – We came back after 26 hours of printing to find this beautiful mess of a failure. This is just a consequence of printing. We kept some of the cooler failures, such as this beauty. Photo image courtesy of Jamber, Inc.

While these sub-sections all start off similarly, the main difference occurs when you expand print quantities or customize the production process. The strengths of more robust traditional manufacturing methods and the strengths of quick, flexible additive manufacturing needs to be considered before starting a project. Many times, the problem solving will have a combination of methods to best fit the needs of the product, budget, and timeframe.

## › 4.3 – What can hobbyists use 3D printers for?

Additive manufacturing's strengths (speed, custom one-off design, and small lot production) are a perfect fit for hobbyist and DIY enthusiast projects. The maker's imagination can sometimes be the limit in creative uses for 3D prints.

Below is a sample of the possibilities:

- **Art** – sculptures, turning mathematical equations into objects, or other artistic creations
- **Custom Hobbies** – model cars, trains, airplanes, table top gaming figures, and rocketry
- **Cosplay** – custom parts for costumes, elements, and props
- **Decorations** – seasonal or everyday household decor
- **Electronics** – custom parts and housings for electronic projects, such as cases, enclosures, mounts, and guides
- **Fashion** – accessories, 3D printed shoes and clothing
- **Gadgets** – custom new designs to make little conveniences such as SD card holders, cell phone holders, and space savers
- **Household Items** – replacements and fixes like the back cover of a remote control or lens cap
- **Jewelry** – make at home or to send to be printed in metal from a service

FIGURE 4.17 – "Prints for fun" – we actually print mini Ultimaker robots for kids whenever we present 3D printing at grade schools. The kids love them. As you can see, sometimes the heads don't print. Photo image courtesy of Jamber, Inc.

- **Learning and Testing** – interworking and gear works, molecules, DNA, typographical maps and other models to help you learn how the real object works
- **Tools** – Functional tools, attachments, add-ons
- **Toys** – from full new toys to add-ons and accessories to existing toys (see **Figure 4.17**)

## Firsthand Account: DIY Hobbyist

### DIY Hobbyist: Lucas Phillips

In this firsthand account, we hear from Lucas Phillips, a DIY hobbyist, on his experiences using a 3D printer for his projects.

INTERVIEW

**LUCAS PHILLIPS**

*Owner, Leyline Studios*

leylinestudios.com

**ST: How did you become interested in using or building a 3D printer?**

LP: My first university job as a researcher exposed me to a lot of emerging technologies. One of the things we got right when they were becoming available was a 3D printer. I became interested in getting my own after I bought my first house and finally had space for hobbies.

**ST: Did you build your own? If so, what type?**

LP: I built a Prusa i3 Mk2 kit.

**ST: If you did build your own, what are some unexpected things you learned during the building process?**

LP: Because the kit itself has a lot of 3D printed parts, I learned a lot about engineering mechanisms to work with 3D printed parts. Getting an in-depth look at how the machine was going together, you see how they do things like route cables and embed bearings.

**ST: How do 3D printers fit into your project(s)?**

LP: I use my 3D printer mainly to create parts for cosplay. I also use it to prototype original designs and create fast MacGyver-esque fixes for things around the house (see Figure 4.18).

**ST: Was your project(s) to solve a problem, art, or for fun?**

LP: Mostly art and fun. Occasionally I solve problems around the house or in the garage with a quick design and print.

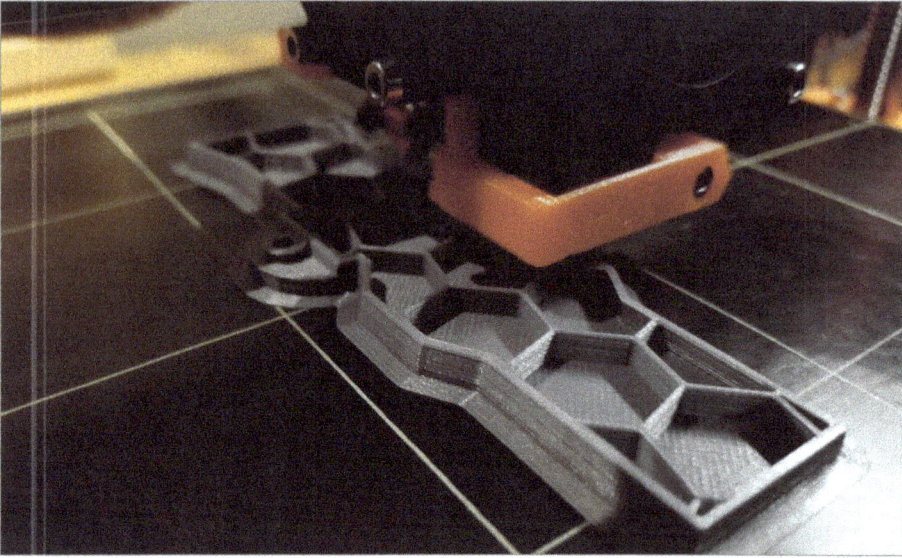

FIGURE 4.18 – A prototype print. Lucas Phillips.

**ST: What benefits do you find in using 3D printers as part of your project(s)?**

LP: Bringing things to life rapidly. Being able to think something up, design it, print it, and then test it. All in a matter of a few hours.

**ST: What type of printer(s), software, and material do/did you use?**

LP: I use a normal Fused Deposition Modeling (FDM) printer. I mainly use Polylactic Acid (PLA). I design with 3Ds Max and use the Prusa version of Slic3r to generate the printable files.

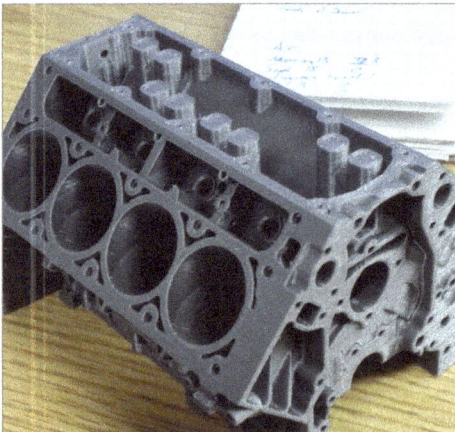

FIGURE 4.19 – Working on a 3D printed part. Lucas Phillips.

**ST: What iterations or changes did/do you go through for an average print?**

LP: Usually subtle size changes or how something fastens to another piece.

**ST: Was there a cost savings versus designing the project in a different method?**

LP: Once I bridged the initial cost of the machine and material, I think there is massive cost savings compared to any other method of prototyping or design (see Figure 4.19).

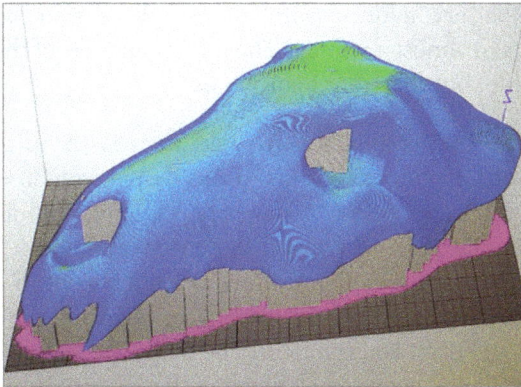

FIGURE 4.20 – CAD 3D model. Lucas Phillips.

FIGURE 4.21 – 3D printed statue that was printed in parts. Lucas Phillips.

**ST: How did you decide to use a 3D printer for your project(s)?**

LP: It just made perfect sense. I have an affinity for 3D and modeling, so being able to "print" my ideas is like a dream come true.

**ST: Did you start with a physical model, a drawing, or a CAD design?**

LP: I usually start off with a rough sketch, then move on to designing it with CAD (see Figure 4.20).

**ST: Do you use 3D scanning?**

LP: I have before. It's not a part of my workflow, but there are definitely benefits to the technology.

**ST: What do you see as the future of 3D printers in your projects?**

LP: I would really like to create prototypes for miniatures and statues (see **Figure 4.21**).

Let's take a closer look at three of these growing areas (household items, electronics, and cosplay) with additional detail and firsthand accounts.

## Household items

One of the broadest uses for DIY makers is household items. When using a 3D printer (or service), you can replace, repair, or design items around the house. Many of these items would be costly to replace or fix.

A home 3D printer (or using a service) allows you to make some interesting household items. Sites like MyMiniFactory and Thingiverse have thousands of items makers can print. From battery holders, switch covers, phone stands, organizers, and replacement parts to almost anything. The quality and use of these online models are highly variable, but there are many gems.

For example, one common household item is the plastic battery cover for remote controls. If the remote you are using breaks or becomes lost, it might be difficult to replace as most manufactures do not sell the cover separately. You could use tape or hope that the batteries do not fall out. If you have a 3D printer or use a printing service, you can print out a replacement cover. You can locate the make and 3D model of the replacement cover needed online. If not, you can find one that is close to what you need and modify it or create your own 3D model.

If you create or modify a 3D model, you can upload it to sites like MyMiniFactory and Thingiverse to help other makers with their prints. This is a big part of the Maker community that we will discuss in Chapter 15.

## Firsthand Account: Household Spray Nozzle

### DIY Hobbyist: Jonathan Torta

I needed to spray paint a model that I was working on. However, the spray paint can I wanted to use has its custom spray nozzle broken (see **Figure 4.22**).

I had replacement nozzles caps, which are cheap and easy to use with different spray patterns (see **Figure 4.23**).

FIGURE 4.22 – A. Broken custom spray paint can nozzle B. Spray paint can stem. Jonathan Torta.

Unfortunately, the spray paint can was of a stem type different from the replacement caps. They were incompatible.

But knowing this, I had an adaptor! (see **Figure 4.24**).

This adaptor allows any of my stemmed caps to attach to a stemmed can rather than cans without stems, as they were made for.

However, the adaptor was made for a smaller diameter can stem! What to do? Order a single quarter-sized piece of plastic from Amazon? No, I will make a 3D printed one!

FIGURE 4.23 – Replacement nozzle cap. Jonathan Torta.

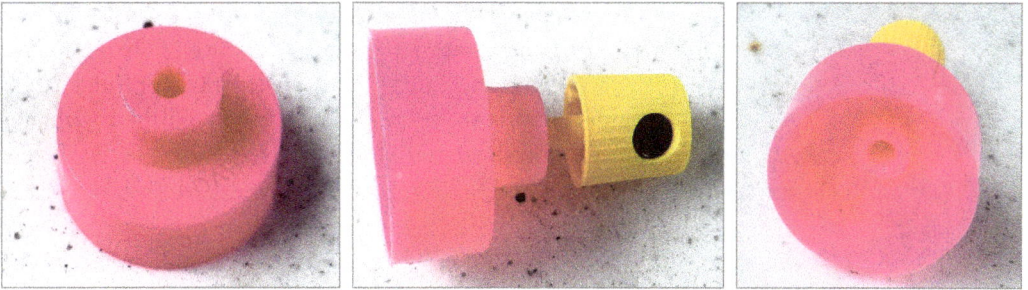

FIGURE 4.24 – Different angles of the replacement nozzle cap and adaptor part. Jonathan Torta.

First, I used the adaptor I had as a template. Functionally, it just had a cylindrical hole through it. The feature is that the holes on each side of the cylinder where different sizes (one for the cap and the other for the can). This adaptor also has a flange to cover the can stem.

I measured the adaptor with a caliper.

A small digital caliper is an excellent and inexpensive tool.

MAKER'S
NOTE

Using the dimensions, I loaded my favorite CAD program for DIY (Fusion 360) and made a copy with the new dimensions of the can stem (see **Figure 4.25**).

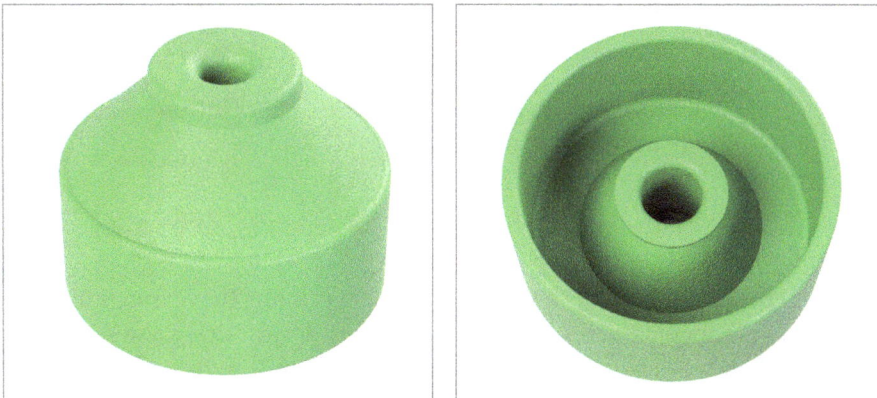

FIGURE 4.25 – Different angles of the CAD adaptor part 3D model. Jonathan Torta.

I made sure to bevel the holes to ensure that no extra material made the holes narrower when printed (which can happen on bottom and top layers). Also, I increased the diameter of the holes by a small amount to allow tight but free insertion of the stems. Holes cannot be the same size as the object you are putting into them.

FIGURE 4.26 – Revised sliced model of adaptor part. Jonathan Torta.

After making the object, I noticed that I would have to use support material because any way I positioned it would have overhangs. This is less than optimal. So, I added bevels to the object that removed the overhangs and added strength to the print (see **Figure 4.26**).

Now to print!

As it is utilitarian object, I do not need thin layers so I printed at 0.3mm first for speed (see **Figure 4.27**).

It took 14 minutes to print and I had my new adaptor! I liked the design; however, I decided to add another bevel on the inside for strength on that side. Another 14 minutes and my final working prototype is done!

FIGURE 4.27 – Different angles of the 3D printed adaptor part and nozzle. Jonathan Torta.

## Electronics

Another growing area for the use of 3D printers is printing the housings for electronics. These include cases, enclosures, mounts, levers, and guides. Custom electronic builds often need custom housing (see **Figure 4.28**).

It can be difficult to find a housing that fits perfectly to the electronic build. For example, a small circuit board that controlled small lights needed to be attached to a set of goggles. After taking measurements of both the circuit board and goggles, a 3D model was made for a custom attachment housing (see **Figure 4.29**).

For the same project, a 3D printed custom ring was printed for inserting and holding the 16 NeoPixel ring in the goggles (see **Figure 4.30**).

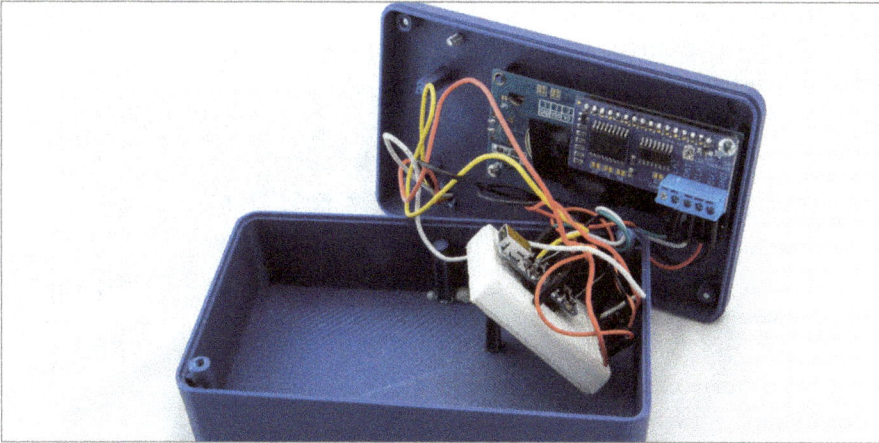

FIGURE 4.28 – Custom 3D printed housing for an electronics build. Jonathan Torta.

FIGURE 4.29 – Custom attachment housing for the microcontroller to a pair of goggles. Jonathan Torta.

FIGURE 4.30 – A 3D printed custom ring for inserting and holding the 16 NeoPixel ring in the goggles. Jonathan Torta.

DIY projects that involve electronics can benefit from custom 3D printed housing to fit the project unique needs.

Makers can also help each other with ideas and prints by posting and searching community sites like Adafuit, MyMiniFactory, and Thingiverse.

**Web link**

A link to Adafruit electronics learning page with a list of their 3D printing projects.

https://learn.adafruit.com/search?q=3dprint#

FIGURE 4.31 – 3D printed classic-looking antique knife switches. Jonathan Torta.

## Cosplay

Cosplay is the act of dressing up as a character from video games, movies, or comics. This can include using supporting props and attending seasonal events.

For example, during Halloween, a mad scientist display had a number of pieces of old brass electronic equipment. These props gave the look of a scientific lab. Some props were found in stores, personal sales, or conventions. However, the real items can be expensive because they are antiques. To save money, 3D printed classic-looking old switches were just one of many props that were built. After they were finished and painted, they looked a lot like the originals (see **Figure 4.31**).

Another example is a 3D printed Egyptian ankh prop that was used

FIGURE 4.32 – A finished and painted 3D printed Egyptian cat figurine. Jonathan Torta.

for another Halloween display. This prop was custom printed to fit the size and detail needed for the display. After finishing and painting, it looks as if it is made of gold (see **Figure 4.32**).

There are a number of online community sites where hobbyists and DIY enthusiasts can learn, collaborate, or help other makers. Some of these sites are specific to certain areas, like cosplay. For example, the site Punished Props Academy has videos, boards, lives stream, projects, and podcasts all geared to help makers with their prints.

A specialized web site for cosplay and prop making.

https://punishedprops.com/3d-printing/

Web link

There are an endless number of projects a hobbyist and DIY enthusiast can 3D print. The ability to custom projects can save the maker time and money and increase usability. And it's fun!

## SUMMARY

Do-it-yourself 3D printing can be divided into two major areas of use (entrepreneurs and hobbyists). There are many questions that need to be answered before knowing if a 3D printing is right for your DIY project. Once these questions have been answered and you decide to use a 3D printer, the results can be well worth the effort especially when considering the cost, flexibility, and speed necessary to complete the project. Many entrepreneurs have started a growing business using their special talents for unique products and many hobbyists are having fun learning how to design a wide range of items.

## APPLYING WHAT YOU'VE LEARNED

1. Continue making your own 3D printer dictionary by adding the definition (in your own words) of five wording relating to 3D printing in this chapter.
2. Explain the two main areas of DIY printing. Which would you like to do and why?
3. What are 10 questions and answers you would need to know before you start your DIY project?

4. Find and interview a DIY person who uses a 3D printer and ask them at least five questions about one of their projects. Share your findings with others.

5. Plan a DIY project you would like to do using a 3D printer and include estimated cost, plus speed information and flexibility information.

6. Summarize one of the interviews in this chapter into a short paragraph.

7. Draw a physical model of a DIY 3D project and explain the reasons why it would be best to use a 3D printer for it.

8. What would you like to happen in the future for DIY users using 3D printers?

9. Explain five things that can go wrong with your DIY project and how you can fix the problem.

10. How are entrepreneurs and hobbyists alike and how they are different?

# Types of 3D Printers

## OVERVIEW AND LEARNING OBJECTIVES

**In this chapter:**

- 5.1 – What are some types of 3D printers?
- 5.2 – What are some industrial additive manufacturing methods?
- 5.3 – Are there common desktop or DIY printers?
- 5.4 – How does an FDM 3D printer work?
- 5.5 – What are some FDM printer components?

## › 5.1 – What are some types of 3D printers?

3D printers come in all shapes and sizes and use different printing methods and materials. In this chapter, we will briefly discuss a few common industrial and desktop 3D printers to give a brief overview of the variety of 3D printers being used today.

- **Industrial** – 3D printers used for industrial reasons. They come in all shapes, sizes, and can use different printing methods. These can be custom-built for a specific job or product.
- **Desktop or DIY printers** – Generally smaller 3D printers that can fit on a tabletop. They can be used for both business, Art and for recreation.

We will concentrate our detailed discussions (starting in Section 5.4) on the Fused Deposition Modeling (FDM) printer as it is the most widely used DIY and desktop printer.

## › 5.2 – What are some industrial additive manufacturing methods?

Industrial printers normally differ from desktop printers in size and scope, although there is some crossover. They also utilize a wider range of additive manufacturing methods and materials.

Some variations include:

- **Primary material used and its state** – for example, plastic or metal as a type, with powder or liquid as the state of the material
- **Fusing method** – using heat or light to melt the material
- **Build structure** – including style of print area, material feeding system, number of axes, and number of print heads and how they function
- **Refinement** – the amount of post-print work that needs to be done with the printed object

Many industrial printers are custom-built for specific tasks. For this discussion, we will list some of the common additive manufacturing methods, selected 3D printers that use the methods, and summarize how they work.

The common additive manufacturing methods and 3D printers include: [1]

# Powder Bed Fusion (PBF)

PBF uses the melting of a powder material (metal or plastic) to fuse particles together. The full or partial melting can be performed by a heated print head, a laser, or electron beam. The build layers are achieved by spreading a thin layer of material to the print chamber area. The print object outline layer is melted. The process repeats until the desired object is made by melting only the object particles in the layers. The un-melted material is then blown away, revealing the new object. This method is typically used for end product manufacturing because if the high strength and the range post-print finishing that can be used. A number of 3D printers use the PBF method.[2]

- **Direct Metal Laser Melting (DMLM)** – uses laser to selectively fully melt powder metal material into liquid pools. These pools fuse to the added material in the next layer. Once the print is finished the unfused powder is swept away, leaving the printed object. This process is strong and can rival traditional methods.
- **Direct Metal Laser Sintering (DMLS)** – uses a laser to selectively sinter/partially melt metal powder layer-upon-layer. It can use a variety of metal powders, like titanium and aluminum.
- **Electron Beam Melting (EBM)** – uses a focused electron beam in a vacuum print area to melt powder metals such as titanium, stainless steel, and copper. It is faster than other PBF methods with larger layers and rougher surfaces that have less residual stress and distortion.
- **Selective Heat Sintering (SHS)** – uses thermoplastic powders selectively melted by a heated print head. New powder is added to the print area and the process is repeated layer-by-layer. This process uses lower temperatures than other PBFs, but also sometimes requires supports for the layers.
- **Selective Laser Melting (SLM)** – uses a laser to selectively fully melt and fuse powder layers in the print area to create an object.
- **Selective Laser Sintering (SLS)** – uses a laser to sinter/partially melt a variety of materials including thermoplastic, glass, or ceramic powder. The powder is added and selectively melted to build upon the previous layer to form the 3D object.

---

### BRIEF REVIEW OF TERMS

**SINTER** – A process of using heat or pressure to form or compact melted material without liquefying.

## Binder Jetting

Binder jetting is similar to an inkjet printer in how it uses nozzles to drop material onto the print area. This method uses a binder material dropped onto a layer of powdered material (metal, glass, or ceramic). After the powdered material is bound, a new layer is ready to build upon the previous layer. This method need finishing after the print is completed and is sometimes brittle overall. This makes it good for more artistic purposes.

## Material Jetting

Material jetting is similar to an inkjet printer. For example, small nozzles deliver drops of waxy photopolymer layer-by-layer onto the print bed to form an object. Ultraviolet (UV) light is used to cure/harden the material before the next layer is delivered. This method uses supports during the process. These supports are deployed using a second set of nozzles. The supports are dissolved after printing is complete. This method is very precise and can print in multiple colors. It is used for prints that need high accuracy and a smooth finish.

- **Nanoparticle Jetting (NPJ)** – uses liquid infused with metal particles. This liquid is deposited onto the print bed in a heated build area. The heat evaporates the liquid leaving a layer of metal

## Directed Energy Deposition (DED)

DED (also known as Direct Metal Deposition) uses thermal energy to melt and fuse material (metal powder and wire filament) within the heated or vacuumed print area. Build platforms and material dispensing vary depending on the thermal energy being used. Lasers, electron beams, and plasma arcs are used to focus the thermal energy sometimes along multiple axes to form the objects. This method is used mostly to print original or repair existing objects.

- **Laser Engineered Net Shape (LENS)** – uses lasers to melt selective area of powdered material located in the print area. The layer then solidifies before the next layer is added.
- **Electron Beam Additive Melting (EBAM)** – uses an electron beam and a vacuum build area to melt metal power or wire filament.

## Material Extrusion

Material extrusion uses thermoplastic filament that is heated and extruded through a nozzle onto the print area in layers. These layers fuse with the previous layer during cooling. This method uses a variety of materials, including plastics and plastics infused with additional materials; however, they can be weaker than other additive manufacturing methods. Because of the overall speed and cost effectiveness of the method, it is ideal for rapid prototyping.

- **Fused deposition modeling (FDM)/Fused Filament Fabrication (FFF)** – this type of 3D printer is one of the more commonly known by the public. It is widely used by both industry and DIY enthusiasts.

## Sheet Lamination

Sheet Lamination uses very thin layers of material bonded to one another. This is achieved by alternating layers of material and adhesive. A variety of materials can be used like paper and metal. Once the layers are set, they are cut by laser or blade to form a print.

## Vat Photopolymerization

Vat photopolymerization uses liquid photopolymer resins in a variety of deployment methods that use light (not heat) to fuse/cure the resins in layers to form the object. For example, one deployment method uses a build area inside a vat of liquid resin. A laser traces the object outline. The laser light solidifies the resin layer-by-layer as it outlines the object. This method is used for printing objects with small details and smooth surfaces.

- **Stereolithography (SL)** – uses a print area vat filled with a liquid photopolymer resin. The resin is cured by an ultraviolet (UV) light layer-by-layer to form the object. This process is becoming popular in DIY 3D printing.
- **Digital Light Processing (DLP)** – is a variant of the SL process that uses mirrors to reflect the UV light. This process is becoming popular to in DIY 3D printing.

## A look to the future

Technology for methods of additive manufacturing and usable materials keeps evolving. New or modified methods are being developed to fit growing applications and needs. New usable materials help accelerate the expansion into new areas of use.

An example is the new experimental process called Rapid Liquid Printing.[3] This print process works within a gel suspension. This allows the ability to physically draw the object from multiple sides in a large gel 3D space. Materials like rubber and foam can be used in addition to types of plastics.

Another example is a custom 3D printer that can print 10 different materials at the same time in a single print. This multi-material 3D printing can also print with different types of materials like plastic and metal, making it easier to create complex objects.[4]

## › 5.3 – Are there common desktop or DIY printers?

While Fused Deposition Modeling (FDM) is the most common desktop printer, Digital Light Processing (DLP) and Stereolithography (SL) 3D printers are quickly evolving in both home and business markets.

Both SL and DLP use light-sensitive resin mediums curable by ultraviolet light. This ultraviolet light is focused onto the surface of a vat filled with a liquid photopolymer resin, creating each layer of the desired 3D object. This process successively prints thin layers over and over resulting in a 3D object.

These are rather new into the DIY and small business market as the resin and printers were expensive and difficult or messy to use. While still messy, the prices have come down and small units are now commonly used for small figurines and dental appliances.

### Fused Deposition Modeling (FDM)

The FDM printer is the most common desktop DIY printer. It prints with melted plastic filament in layers (by the material extrusion method). We will focus on this type of printer and its inner workings in Section 5.4 of this chapter

### Stereolithography (SL)

Stereolithography (SL), also known as Stereolithography Apparatus (SLA) (a registered trademark of 3D Systems), is just starting to make it to the high-end DIY market. SL is a vat photopolymerization method that scans a laser to activate the resin (i.e., draws a line like a small laser show). These tend to make very smooth outlines but are slower than DLP and may have issues with some details, like hard corners.

### Digital Light Processing (DLP)

DLP, a vat photopolymerization method, is a variant of SL and may be the next big thing in high-end home DIY. It works like a projector TV and reflects light with a matrix of tiny mirrors. These tend to be faster than SL, as they do an entire layer (slice) at one time, but are limited to the resolution (size) of the mirrors.

# › 5.4 – How does an FDM 3D printer work?

Effectively, an FDM 3D printer can be envisioned as a hot glue gun with a small nozzle attached to a computer-controlled mechanical carriage that repeatedly outlines an object, building up the height with each successive layer it deposits. There is even a version that is a handheld pen (3D pen) that does not have a carriage, but this is more for improvisational art than printing.

Here is how the FDM 3D printing process works:

1.  A spool of thermoplastic filament is loaded into the printer. The filament is fed into the extruder. The nozzle temperature is increased to the desired level. The filament in the nozzle melts.

2.  The extruder is attached to a 3-axis system that allows it to move precisely and reliably in the X, Y, and Z directions (X – right–left, Y – forward–backward, Z – up–down). The melted filament/material is extruded in thin strands. This is deposited layer-by-layer, outlining the object being printed, where it cools and solidifies. The cooling of the extruded filament can be accelerated with the use of a cooling fan attached on the extruder.

3.  Often just an outline is not enough and the object needs some infill to hold up upper layers and to keep the object strong. Generally, this infill is a loose grid of lines, similar to cross-shading or coloring in a shape after you draw the outline (See Chapter 7).

4.  After a layer is finished, the extruder or build platform is moved to make room to deposit the next layer.

5.  This process repeats until the object is complete.

## Types of FDM 3D printers

There are a few main types and configurations of FDM printers.

- Freehand pen
- Cartesian
- Delta
- Polar
- Robotic arm

These are delineated mostly in the type of coordinates system the printer uses and how the print head moves around the print area (see **Figure 5.1**).

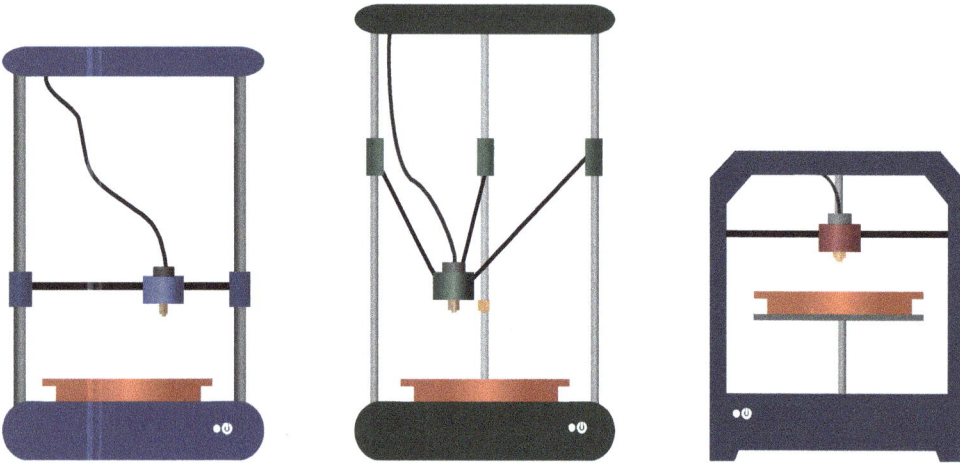

FIGURE 5.1 –The most common FDM printers from left to right: polar printers, delta, and Cartesian. Michelle McGough.

Let's take a closer look at these types of 3D printers.

## Freehand Pen

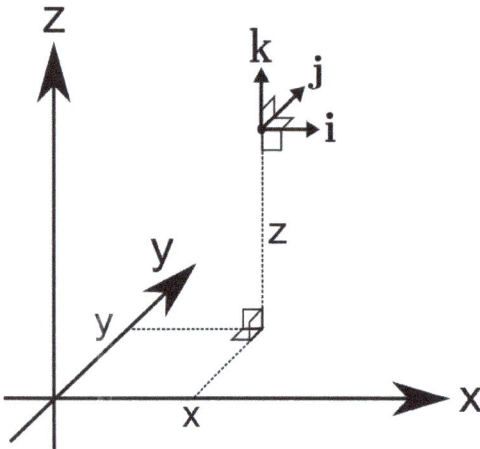

FIGURE 5.2 – A depiction of Cartesian coordinates (xyz) with the elementary basis vectors (ijk). Math buff.

A freehand pen is a 3D printer without the carriage. You become the carriage. These pens are great for doodling and for arts and crafts. You can also make 3D designs and structures freehand by creating a base and making a framework.

## Cartesian 3D FDM Printers

Cartesian 3D printers are by far the most common printer available. They are based on the Cartesian coordinates system in mathematics using three axes: X, Y, and Z to move and position the print head (see **Figure 5.2**).

Often these printers move the bed only on

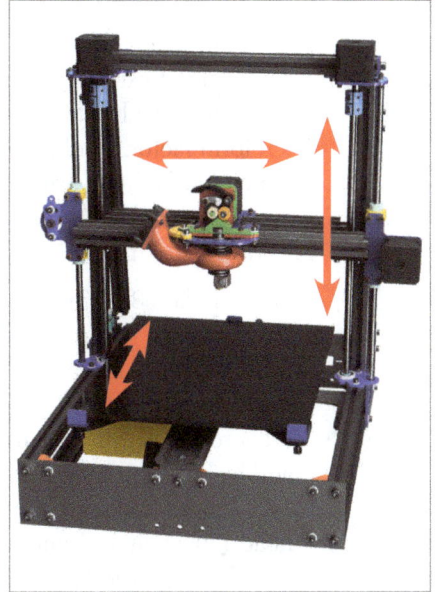

FIGURE 5.3 – The two most common Cartesian printer configurations. Jonathan Torta.

the z-axis, with the print working two-dimensionally on an x-y plane. Some versions, however, move the print head in the x and z and the bed moves in the y (see **Figure 5.3**).

These printers are currently the most developed and versatile; some even have different heads for laser cutting and computer numerical control (CNC) machine tools.

## Delta 3D FDM Printers

Delta printers are a newer type that utilizes the same coordinates system but moves the print head in a unique way. This involves a round printing bed combined with an extruder that is attached at three triangular points. Each of the three points then moves up and down on tracks, determining the position and direction of the print bed. Deltas move the print head exclusively (see **Figure 5.4**).

Delta printers can be quite fast, but are potentially

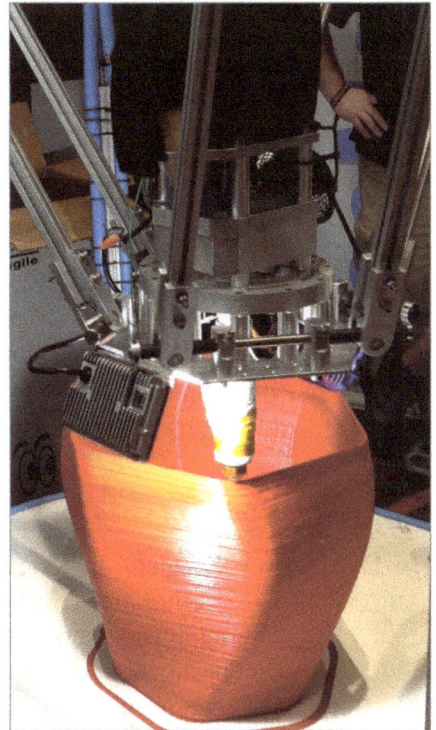

FIGURE 5.4 – A large delta-style 3D printer by SeeMeCNC capable of printing an object up to 4 feet in diameter and 10 feet in height. Z22.

less accurate than a conventional Cartesian printer. Also, to reduce the weight of the print head, most use a Bowden tube to feed the filament to the print head. This style of extrusion potentially limits the types of filaments that can be used and can cause binding within the tube itself. Due to its round build plate and tall form factor, a delta printer may be more appropriate for specific build sizes, depending on your print.

## Polar 3D FDM Printers

With polar 3D printers, we have another coordinates system, as the head positioning is not determined by the x, y, and z coordinates, but by an angle and length. This means that the bed rotates (R) and the print head moves in and out and up and down (X and Z) as seen in **Figure 5.5**.

One advantage of polar printers is they only utilize two motors to move to the required coordinates, saving money and parts.

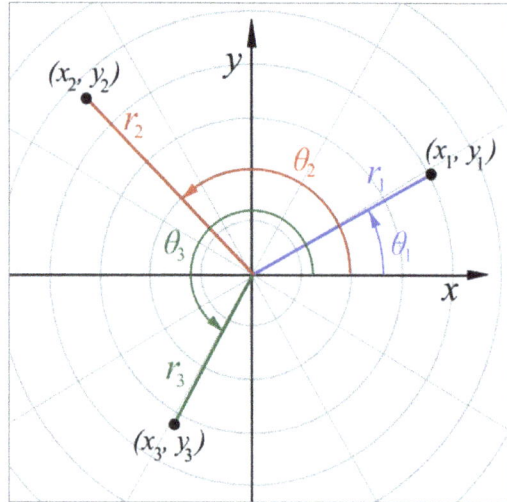

FIGURE 5.5 – Polar coordinates system. Svjo.

## Robotic Arm 3D FDM Printers

Lastly, there are robotic arm-based 3D printers. These can allow more flexibility, mobility, and size and thus can be scaled and positioned more accurately than the other printers, making remote or large works possible. However, they cannot move as quickly or as accurately.

## › 5.5 – What are some FDM printer components?

Even though the location and number of parts might change from one 3D printer to another, their overall functions remain the same. Understanding how a 3D printer works and how each part affects the print can help you maintain the functionality of your printer or troubleshoot issues that might arise during the printing process.

Some differences include: the style of drive (direct drive or belts), heated print beds, enclosed build areas for heat management, multiple print heads, moving print beds, removable print beds, additional controls, liquid-crystal display (LCD) screens, touchscreens, and Wi-Fi, to name a few.

For example, additional print heads allow the use of more than one filament at the same time. This is often used to print a different material for support.

Another example is that the most current printers can either automatically load new filaments into the head as needed or mix many different filaments at the same time. This allows two or more colors and even blending.

See **Figure 5.6** for a general view of how these main parts fit into a common FDM 3D printer build.

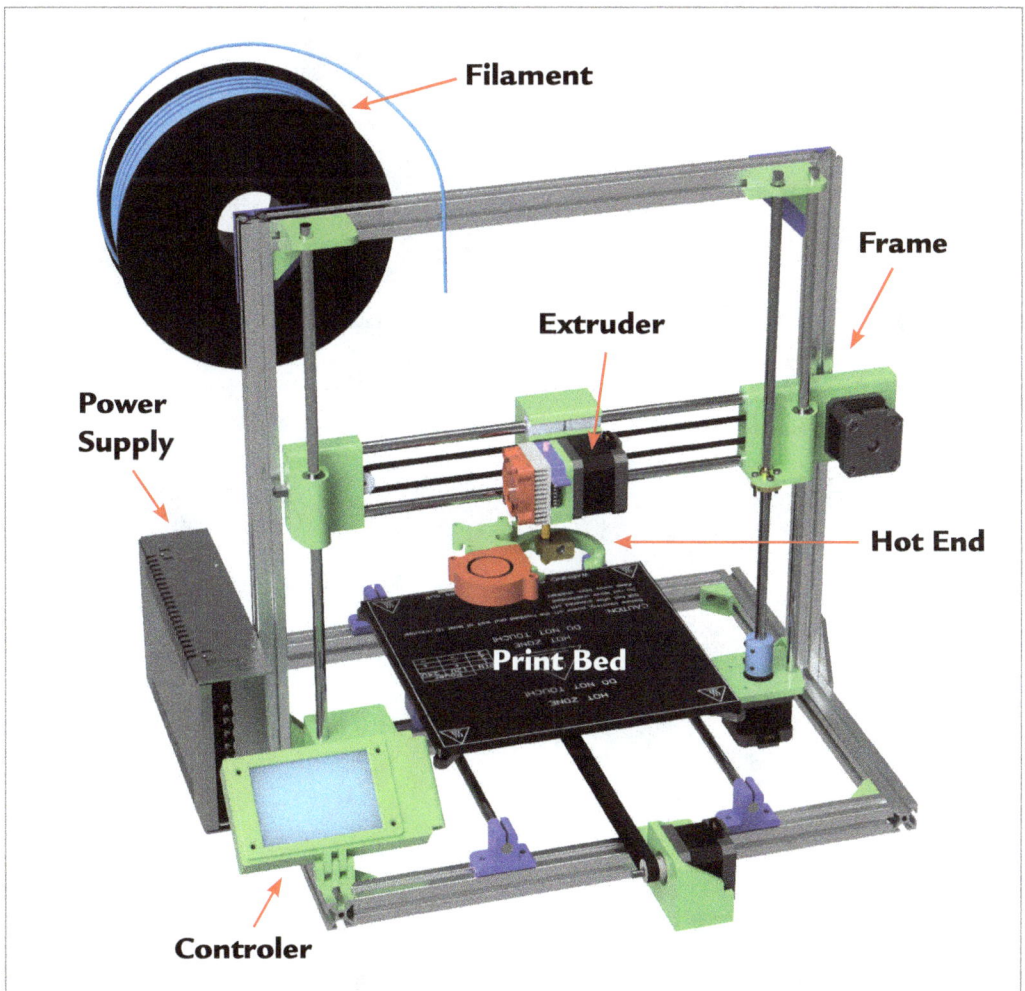

FIGURE 5.6 – 3D printer with parts labeled. Jonathan Torta.

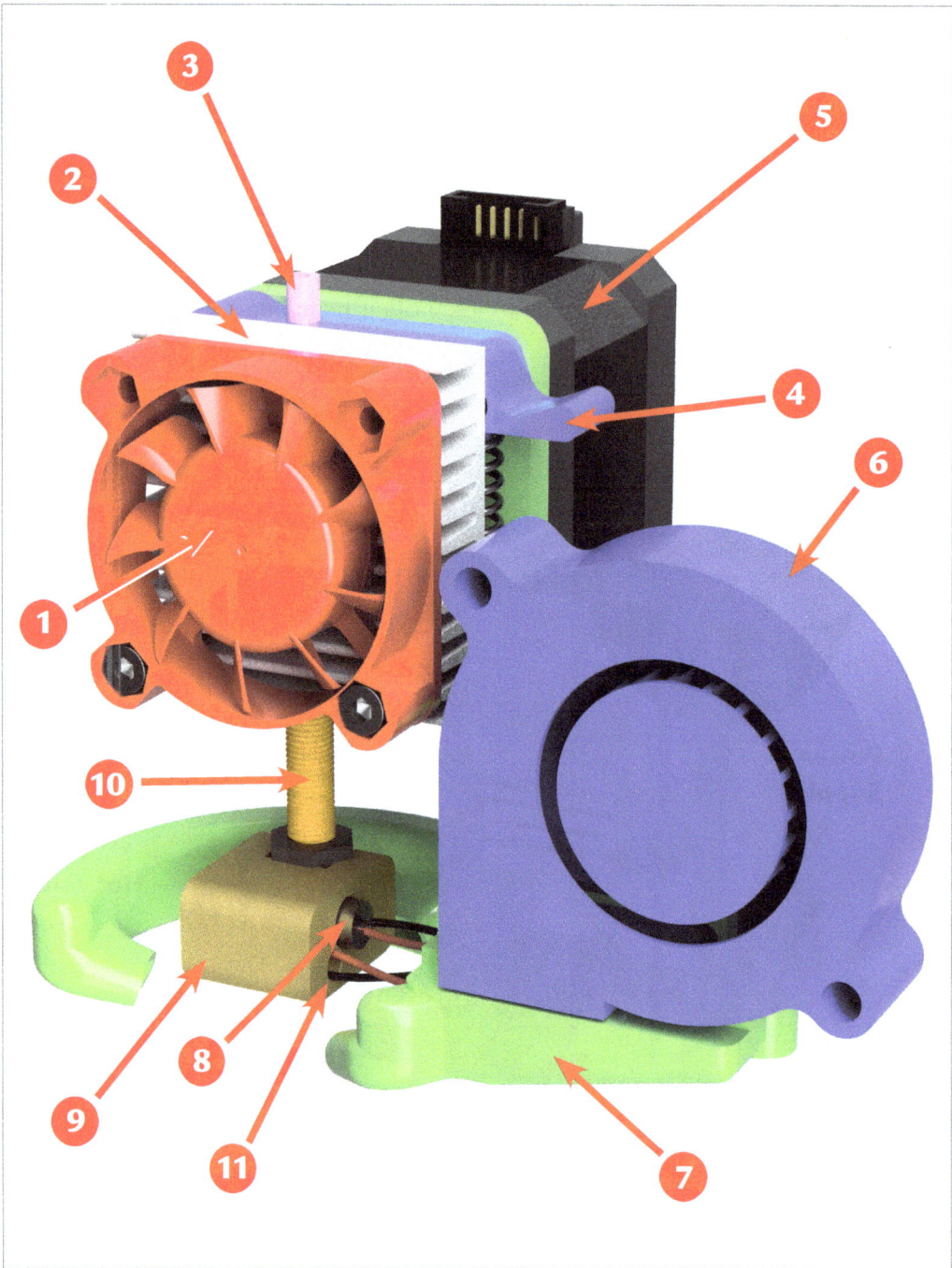

FIGURE 5.7 – 1. cold end fan, 2. heatsink, 3. guide tube, 4. tension lever, 5. stepper motor, 6. filament cooler blower fan, 7. fan duct, 8. heater cartridge 9. heater block, 10. heat break, 11. thermistor. Jonathan Torta.

Depending on the type of 3D printer, the parts and their locations may differ from the diagram.

MAKER'S
NOTE

In the next section, we will highlight the common parts of a typical FDM printer divided into three main areas, including mechanical, bed/motor/frame, and electrical components.

## Mechanical Components

The main component of any FDM 3D printer is the extruder. The extruder pulls, melts, and extrudes a thin filament of plastic through a nozzle. It functions like an automatic hot glue gun. (See **Figure 5.7**.)

The extruder consists of two parts:

- **The cold end** draws the filament in from the spool and pushes it through to the hot end
- **The hot end** melts the filament with a heating element and extrudes it out through a nozzle

### Cold End

The cool end consists of mainly the filament drive. Below is a list of the cold end's subcomponents and their functions within the extruder.

#### Hobbed Gear

This is half of the filament drive in the cold end. This has teeth and grips into the filament creating traction.

#### Idler Gear

The idler gear is a spring-loaded wheel that provides constant pressure against the hobbed gear to pinch the filament between. This allows constant pressure to be placed on the filament for consistent movement, even when the diameter changes slightly. Typically, the tension on the idler is adjustable, so it squeezes the filament neither too tightly or too loosely. (See **Figure 5.8**.)

#### Direct Drive and Bowden Drives

The two main variants of the filament drives are direct drives and Bowden drives. These are differentiated by how the cold end handles and moves the filament. With direct drive, the cold end with the filament drive is right above the hot end and they both move as a unit. The filament is driven from the spool through the cold end and directly into the hot end.

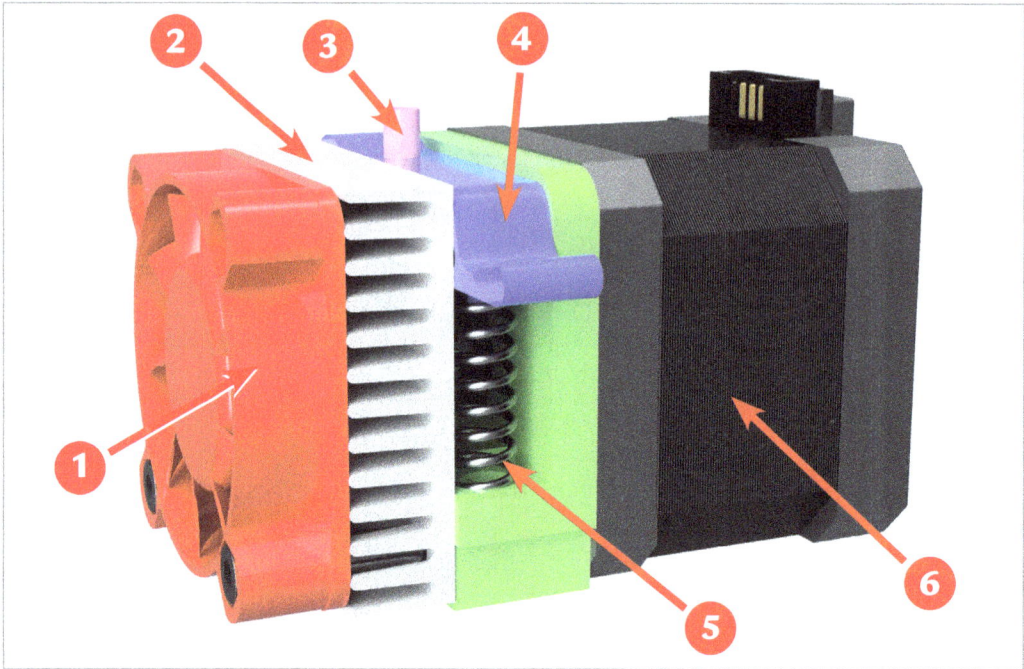

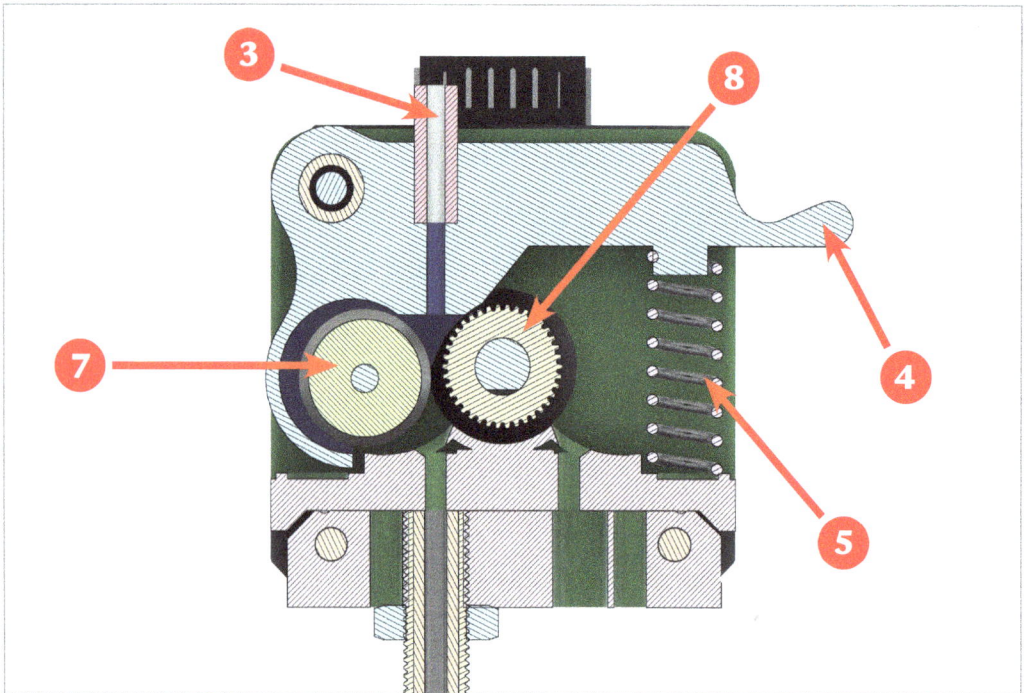

FIGURE 5.8 – 1. fan, 2. heatsink, 3. guide tube, 4. Tension lever, 5. tention spring, 6. stepper motor, 7. idler gear, 8. hobbed gear. Jonathan Torta.

In a Bowden drive, the cold end is separated and stationary, connected by a long tube (called a Bowden tube) to the hot end. The filament is driven from the spool through the cold end and directly into the hot end. (see **Figure 5.9.**)

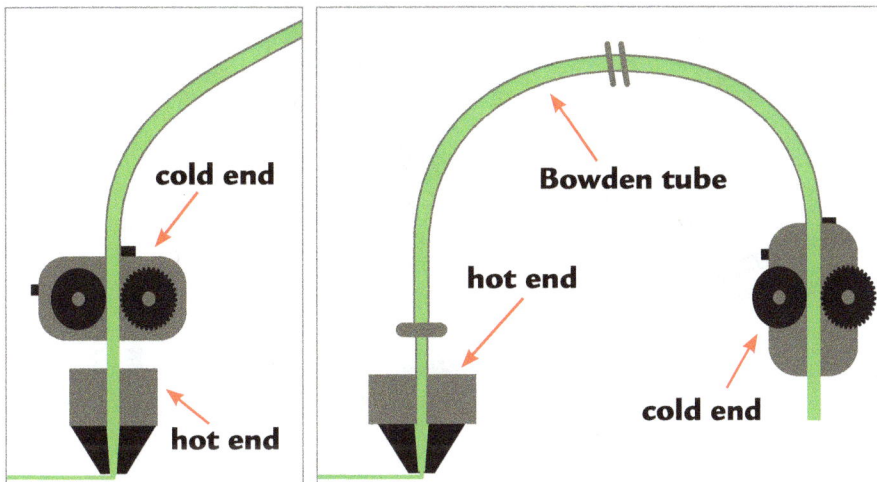

FIGURE 5.9 – Direct Drive vs. Bowden. Michelle McGough.

## Hot End

This consists of all the parts used to heat and extrude the material. Below is a list of the hot end's subcomponents and their functions within the extruder (see **Figure 5.10**).

### PEEK/PTFE or All-Metal Insert

The hot end has two main variants:

- One version has a PEEK/PTFE liner that guides the filament through the heat break into the nozzle. This makes it easier to print certain materials, like PLA. It helps with friction and reducing heat creep into the cold end. The liners are ideal for lower-temperature materials like PLA, though higher temperatures can degrade the PEEK/PTFE tube over time and cause failures.
- An all-metal version of the hot ends is able to reach much higher temperatures and print a wider range of materials. It tends to be more expensive and need robust cooling from a heatsink and fan to stop heat from moving up to the cold end.

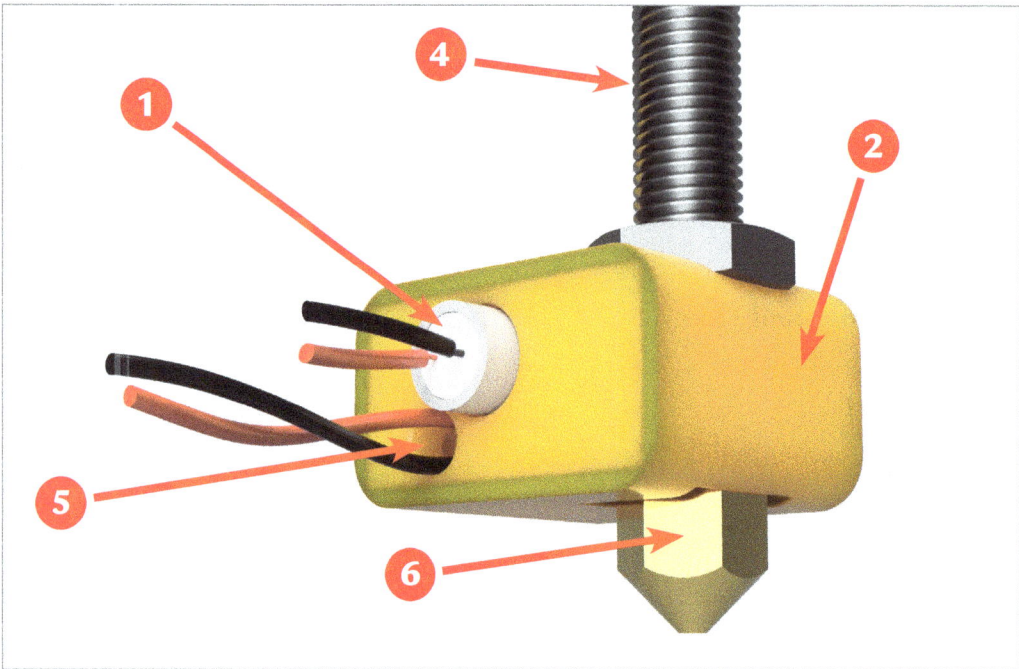

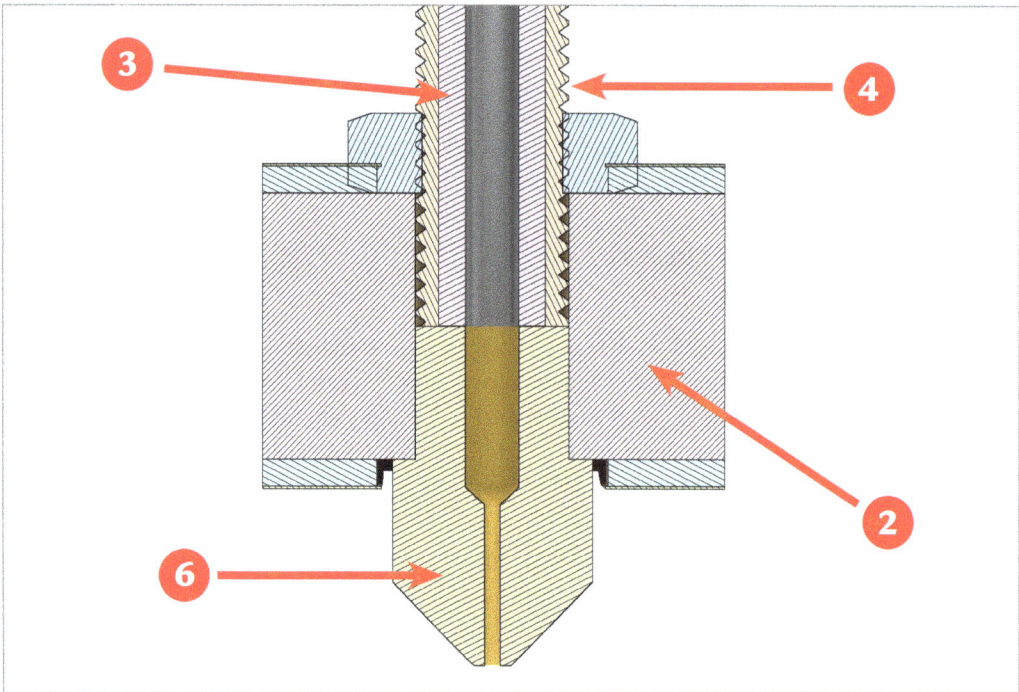

FIGURE 5.10 – **Top:** hot end exterior 1. heater cartridge 2. heater block, 4. heat break, 5. thermistor, 6. nozzle. **Bottom:** hot end cross-section 2. heater block, 3. PEEK/PTFE insert, 4. heat break, 6. nozzle. Jonathan Torta.

## Heater Cartridge

The heater cartridge is responsible for heating up the hot end. The heat block usually clamps around the heater cartridge to provide good contact. Most hot ends use a ceramic heater cartridge, though some older designs use power resistors or nichrome wire.

## Heater Block

Usually made from aluminum but can be copper plated. The heater block joins the nozzle to the heat break and holds the heater cartridge and thermistor.

## Thermistor/Thermocouple

This part senses the temperature of the heat block. There are various types of sensors for determining the temperature of the hot end. They are essentially electronic thermometers. Some look like a small glass bead or small metal cylinder with two wires attached (which are typically insulated with glass fiber or Teflon). For high-temperature printing, a thermocouple may be used instead of a thermistor.

## Heat Break

The heat break is the interface where hot meets cold. It is usually a thin tube, made of stainless steel for its low thermal conductivity. The transition is made to be as short as possible so only a small amount of filament is in a molten state. It also connects the heat block to the heat sink.

## Heat Sink/Hot End Fan

A heat sink is a passive heat exchanger that transfers the heat generated in an area to a fluid medium –in this case, air – where it is dissipated away. The heatsink on the hot end transfers heat away from the top of the hot end and ensures that heat does not travel too far up the plastic and melt it prematurely before it reaches the hot end. This is called heat creep; too much creep can cause jams. Often a fan is used to help the heat sink transfer heat by moving an increased volume of air past the heat sink, thus making it much more efficient.

## Layer Cooling Fan(s)

This fan's purpose is to cool the material immediately after it is extruded from the nozzle, quickly hardening it and helping the freshly laid filament hold its shape. The slicer can ramp the speed of the fans or turn them on and off as needed, depending on the material type and need. This is particularly useful for bridging and thin towers that need to be cooled quickly to hold shape.

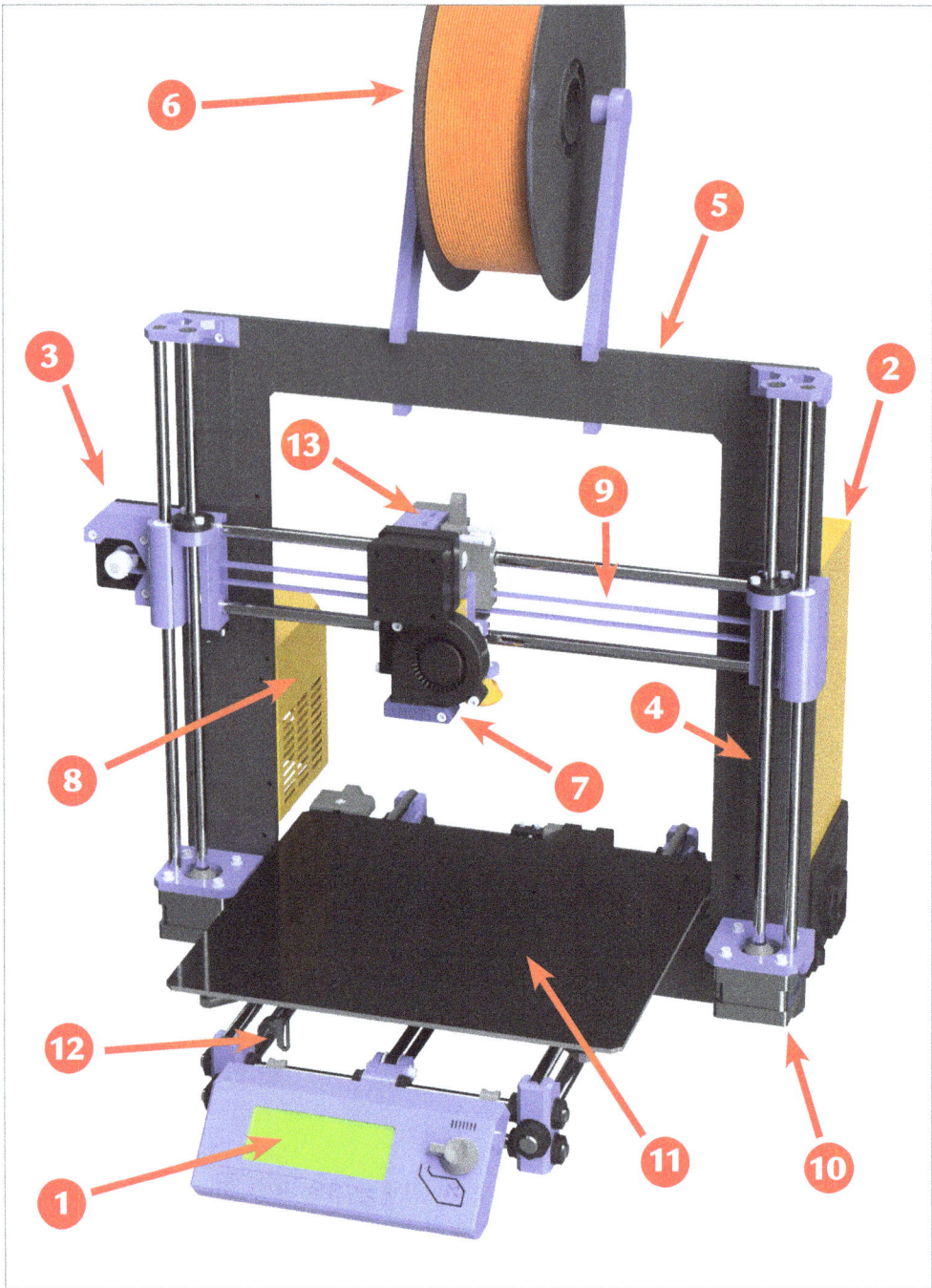

FIGURE 5.12 – 1. display, 2. controller board, 3. "x" axis stepper motor, 4. "z" threaded rod, 5. frame, 6. filament spool, 7. hot end, 8. power supply, 9. pulley, 10. "Z" axis stepper motor, 11. print bed, 12. "y" axis rod, 13. extruder. Jonathan Torta.

### Nozzle

The nozzle takes the molten filament (typically 1.75–3mm) and tapers this down to the nozzle size (typically around 0.4mm). These are often made of brass for its good heat conductivity. Sometimes they are hardened stainless steel and tungsten coated. Even new ruby-tipped nozzles can be used for very abrasive materials, as brass is not suitable for printing abrasive materials (such as glow in the dark and metal-filled filaments) and will wear quickly. Nozzles are interchangeable and come in various sizes. You might use a smaller nozzle for finer detail or a larger nozzle to print faster.

## Bed, Motor, and Frame Components

Below is a list of the components comprising the printing bed, stepper motors, frame, and other mechanisms and their functions within the 3D printer (see **Figure 5.12**).

### Print Bed

The print bed is the surface that your objects are printed onto. It is commonly an aluminum base with a variety of different bed surfaces and an optional heating element. The heating element can help various materials stick to the platform better and also regulates the temperature of the object to limit warping from cooling too fast.

### Bed Surfaces

The bed surface helps the material stick to the underlying print bed during printing process and some allows it to be removed more easily when printing is done. Some surfaces consist of Kapton Tape, painters' tape, borosilicate glass, PEI Sheets, and Flex Plates.

### Bed Leveling and Tramming Mechanisms

Printers need a way to *tram* (or adjust) the print bed to be perpendicular to the print head. Often, multiple spring-loaded screws are used to adjust the bed. This is technically called tramming, not leveling, as it is not leveling the bed to the ground. Some printers have automatic systems that tram or additionally map out minute imperfections on the print bed to allow the printer to handle a slightly uneven surface.

### End Stops (one for each axis)

The end stops are a switch that is triggered when a moving part moves into it.

Most are mechanical, but others can be optical. They are a simple method for the printer to determine where the limit of an axis is. This is also how the printer "homes" itself (calibrating its position after being moved or off) and finds its starting point before printing.

### Threaded Rods or Leadscrews

These are most often used with the printer's z-axis. They are adjustable, forcing a nut attached to the print bed to move up and down. This in turn moves the print bed.

### Stepper Motors

Stepper motors are special brushless DC motors that achieve a high level of precision in small movements and can rotate in increments giving them precise control over their position and speed. These are used to move the all parts of the printer attaching to gears, belts and screws. Most printers use NEMA 17 type motors (1.8 degree 200 steps-per-revolution).

### Belts

Belts allow the stepper motors to be placed in better locations and transfer the rotation of the motor to another location/shaft or to create linear motion. The most commonly found belts for 3D printers are made of rubber, some reinforced with fiber glass. Rubber belts have very good flexibility, but do stretch over time; tensioners are required and must be periodically readjusted. Commonly, the stepper motors have toothed sprockets that drive the belts; a corresponding toothed belt should be used.

### Frame

The frame holds all the parts together and is the base skeleton of the printer. Printer frames can be made of laser-cut plywood, sheet metal, aluminum beams, plastic, or a combination. Many parts can be 3D printed. It is best to have the frame as rigid as possible, as any flexibility can create flaws in the print.

### Enclosure

Having an enclosure for your 3D printer can be handy, as most printers are just bare frames. Enclosures control the environment, keeping the interior clean and aiding in the safe operation of the printer. If you do not have an enclosure, it is easy to construct your own. Something as simple as a cardboard box could suffice but they can also range up to fancy DIY Plexiglas with 3D printed hinges and doors.

## Filament and Holder

This is the actual plastic material the printer makes the object out of. The two main sizes of filament are 1.75 mm and 3 mm. There are a variety of different materials (see Chapter 6). Most often can be purchased in spools.

# Electrical Components

## Power Supply

A power supply takes the AC electricity from the wall and converts it to low voltage DC power for use by your printer. Some printers run 12-volt systems while others run 24-volt systems.

## Motherboard/Controller Board

The motherboard is the brains of the printer. It takes the G-Code commands given to it and orchestrates their running. The motherboard contains a microcontroller, effectively a small computer. It may also contain all the circuitry (stepper drivers) needed for the motors, reading the sensors, and communications to your computer and user interface. (See **Figure 5.13**.)

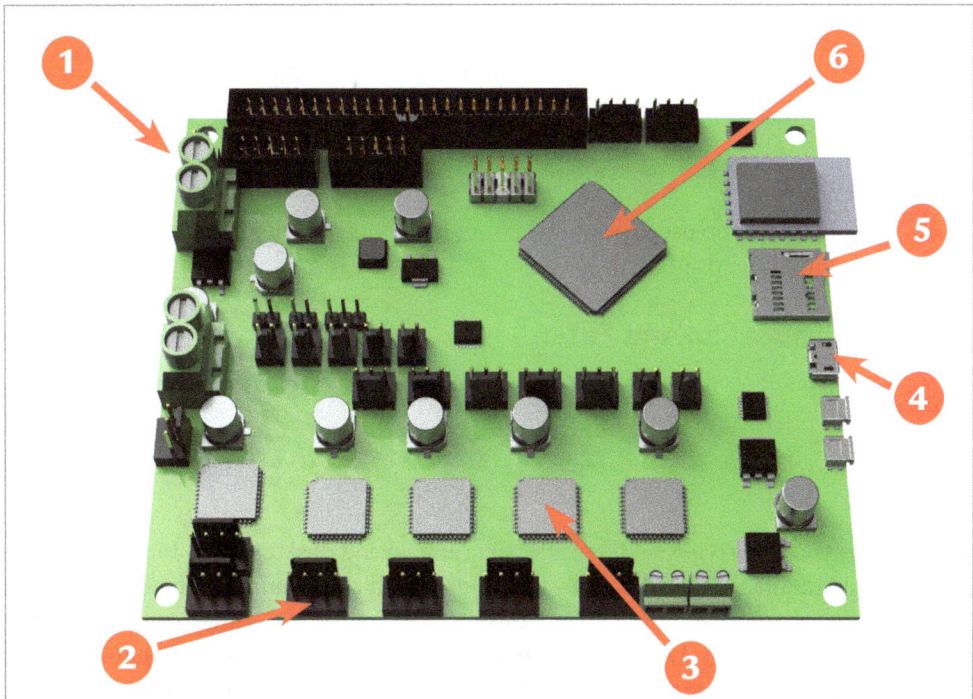

FIGURE 5.13 – 3D rendering of a motherboard. 1. heated bed power, 2. stepper motor connectors, 3. stepper drivers, 4. USB connector, 5. SD card slot, 6. microcontroller. Jonathan Torta.

### SD Card Slot

Most printers have an SD card slot from which they can load G-Code files. This allows them to run independently without a computer.

### USB Connector

Most printers have a USB connection from which they can stream G-Code files, control the printer, and update the firmware when connected to a computer or controller.

### Stepper Drivers

These drivers are responsible for running the stepper motors and controlling their position and the amount of electrical current fed to the motors. Many motherboards have the stepper drivers built in or in separate modules.

### Screens and User Interfaces

Some printers have LCD screens so they can be controlled directly without hooking them up to a computer. These can be basic text displays or advanced touchscreens as a local way to control your printer.

### Lights (LEDS)

These are to light the workspace and some can be fun, changing colors when print is done.

### Camera

A camera can be used for photos, time-lapse video of the print, or remote viewing during the printing process.

### Enclosure Fan and Filter

Some encloses also have an exhaust fan with filter to clean the air.

### SUMMARY

There are two types of 3D printers: industrial and desktop/DIY. Industrial Additive Manufacturing utilizes a wide range of additive manufacturing methods and materials such as Powder Bed Fusion (PBF), binder jetting, material jetting, Directed Energy Deposition (DED), material extrusion, sheet lamination, and vat photopolymerization. The common desktop or DIY 3D printers use three common 3D technologies: Fused Deposition Modeling (FDM), Digital Light Processing (DLP), and Stereolithography (SL). There are five main types and configurations of FDM 3D printers depending on the coordinates the system uses and how the printer heads move. FDM 3D printer components depend on the make, style, and modifications in both the mechanical and electrical components.

## APPLYING WHAT YOU'VE LEARNED

1. Continue making your own 3D dictionary by adding the definition (in your own words) of five words related to 3D printing in this chapter.

2. How are the industrial and desktop/DIY 3D printers the same and different?

3. In your own words, summarize Powder Bed Fusion (PBF).

4. Compare and contrast the binder jetting and material jetting methods.

5. Compare and contrast the Directed Energy Deposition (DED) and material extrusion methods.

6. Explain how the five main types of FDM printers work, in your own words.

7. If you have an FDM printer, what type is it and how does it work?

8. Discuss the extruder's two mechanical parts (cold end and hot end) and how they work.

9. Explain four mechanical components and four electrical components of an FDM 3D printer.

## REFERENCES

[1] – https://www.ge.com/additive/additive-manufacturing/information/additive-manufacturing-processes

[2] – https://www.ge.com/additive/additive-manufacturing/information/powder-bed-fusion

[3] – https://selfassemblylab.mit.edu/rapid-liquid-printing/

[4] – https://www.cnet.com/news/cutting-edge-3d-printer-prints-in-10-materials-simultaneously/

By Jonathan Torta

# Printable Materials including FDM Filament

## OVERVIEW AND LEARNING OBJECTIVES

**In this chapter:**

- 6.1 – What are some materials 3D printers can use?
- 6.2 - How do I select the materials to use for my FDM printer?
- 6.3 – What are some common basic filament for FDM printers?
- 6.4 – Are there hybrid materials?
- 6.5 – What specialty materials can I print with my FDM printer?

## › 6.1 – What are some materials 3D printers can use?

3D printer materials come in a wide range of types, color, and styles. Each has advantages and disadvantages that should be considered before selecting the material to best fit the project needs.

The usable material will also depend on the selected 3D printer. For example, some industrial printers use powders or resin, while common desktop 3D printers use types of plastics.

The following is a list of common printable materials used by 3D printers:

- Carbon Fiber
- Ceramic
- Edible
- Glass
- Metal
- Nylon
- Paper
- Plastic
- Wood

Some materials, like plastics, have different properties that can greatly affect the final print. For example, there are materials that are conductive, flexible, color-changing, glow-in-the-dark, or have magnetic properties.[1] See **Figure 6.1** for an example of a glow-in-the-dark print.

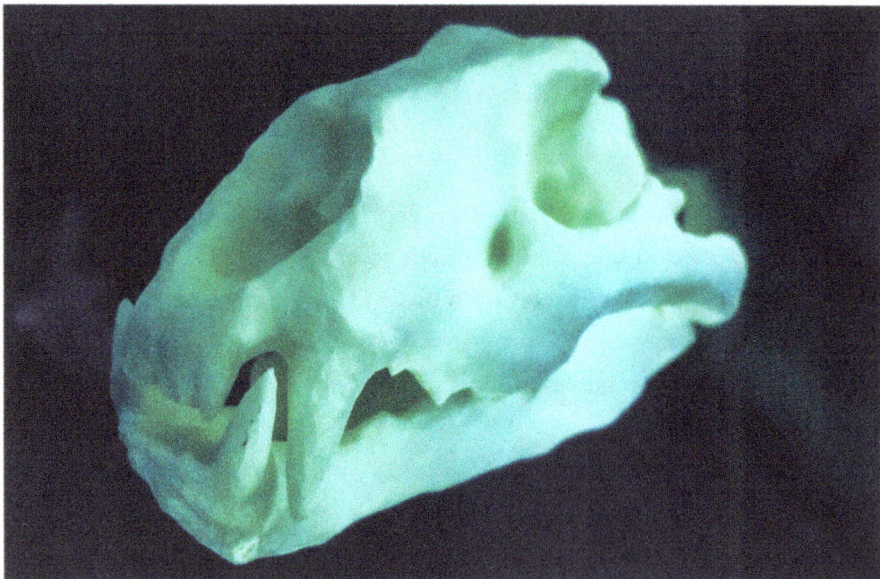

FIGURE 6.1 – A glow-in-the-dark print. Jonathan Torta.

*Color images and interactive links are located in extras files and on the DVD.*

IN EXTRAS

This chapter will highlight the materials appropriate to FDM printers because they are the most widely used by DIY makers.

## › 6.2 – How do I select the materials to use for my FDM printer?

Thermoplastic filaments are a common material used by FDM printers. When heated, these materials become soft and pliable; when cooled, they become hard. They are packaged in spools that attach to the 3D printer. **Figure 6.2** shows spools of printing materials in a variety of colors and sizes.

FIGURE 6.2 – 3D printing materials. Maurizio Pesce.

While the term materials can cover many different types of printable substances, for the rest of this chapter we will be using it interchangeably with filament because we will be focusing on materials printable by FDM 3D printers.

MAKER'S NOTE

### Things to Consider

Selecting the filament to best fit your printer and your project is an important step in the printing process that directly affects your final print. Every filament has specific advantages and disadvantages. Let's take a look at some of the factors to consider.

## Size

Select the filament size that matches the requirements of your printer. Currently, there are two main filament sizes for desktop 3D printers.

- 1.75 mm
- 3.0 mm

## Type

Plastic is the main material used in FDM prints. However, the plastic can be combined with additional materials, altering its properties and changing the overall appearance and strength of the final print.

- Basic
  - › Raw plastic
  - › Raw plastic plus colorants
- Hybrids
  - › Raw plastic plus additional materials
  - › Raw plastic plus additional mixtures of plastics
- Specialty materials
  - › Different formulations

There can be a tradeoff when selecting a filament. Aesthetic and tactile appeal can come at the cost of reduced flexibility and strength – something to keep in mind when picking the material.

## Cost and Availability

Materials can be purchased from retailers including Amazon, Home Depot, and specialty stores. Depending on type and quantity, the materials can range from $10 to over $100 per spool. Some specialty filament may be more difficult to obtain than the more common basic materials.[2][3]

## Quality

Not all filaments are equal in quality. The quality of your filament dictates the quality of your print. Before you purchase filament, be aware of the following:

- **Diameter and consistency** – the filament needs to be consistent. If the diameter grows too large, it will simply jam the printer; if it is too small, it may no longer feed.

- **Formulation, impurities, viscosity, and debris** – lower-quality filament can introduce impurities and other imperfections that can cause issues when printing.
- **Moisture content** – good packaging is critical, as the plastic will absorb moisture from the atmosphere relative to the humidity and duration of exposure.
- **Potential age issues** – the older the filament is, the greater the potential of compromised storage and degradation.

We will go into more detail about the importance of selecting the best filament to match the needs of your project in Chapter 11 when we talk about calibration.

## › 6.3 – What are some common basic filament for FDM printers?

Before you select a filament to use for your project, there are a number of questions to consider.

- How easy is it to use?
- How will it stick to print bed?
- What temperature does it print at?
- How strong or flexible is it?
- What are the potential problems?

In this section, we list common filament types with information on properties, usage, application, pros and cons, strength, flexibility, durability, printing temperature (general), bed temperature (general), bed adhesion, and additional notes to answer some of these questions.[4]

### Acrylonitrile Butadiene Styrene (ABS)

**Notable Properties**

- Useful for durable parts that need to withstand higher temperatures, easy to print with, strong plastic

**Typical Usage**

- LEGO®, sports equipment, phone accessories, toys, various handles, cases for electronics.

### 3D Print Application

ABS is ideal for moving parts, such as automotive parts, electronic housings, and toys.

### Pros

- Very durable and strong
- Lightweight and slightly flexible
- One of the cheapest thermoplastics on the market
- Tolerates higher temperatures
- One of the first 3D printing materials

### Cons

- Petroleum-based, less-biodegradable plastic
- Requires a higher temperature to reach a melting point
- Creates fumes which may irritate people
- Prone to shrinkage and warping

### Strength/Flexibility/Durability

- *Strength:* High | *Flexibility:* Medium | *Durability:* High

### Printing Temperature (general)

- 210 °C – 250 °C

### Bed Temperature (general)

- 80 °C – 110 °C

### Bed Adhesion

- Kapton Tape/hairspray

### Additional Notes

3D printer enthusiasts should be mindful of the filament's high printing temperature, tendency to warp during cooling, and the intensity of its fumes. Be sure to print with a heating bed and in a well-ventilated space.


## Polylactic Acid (PLA)

### Notable Properties

- Present odor, low-warp, eco-friendlier, less energy to process

### Typical Usage

- Biodegradable medical implants, candy wrappers, food containers, prototype parts, art pieces

### 3D Print Application

PLA has the ability to degrade into inoffensive lactic acid in the body and due to this property is used in medical suturing and some surgical implants. Surgically implanted screws, pins, rods, or mesh naturally break down in the body.

Additionally, PLA is considered food safe. Thus, it is used in food packaging, candy wrappers, disposable tableware, disposable garments, hygiene products, diapers, etc. Various colors and additives may make it less so.

### Pros

- Easiest material to work with
- Works well for beginners
- Less prone to warping compared to ABS
- Available in many variants including full range of colors, translucent, and glow-in-the-dark
- May produce a sweet aroma that smells like candy when heated

### Cons

- Can be strong but brittle
- Attracts water molecules and becomes brittle at times, making it difficult to print if not stored correctly
- Water saturated PLA needs a higher extrusion temperature

### Strength/Flexibility/Durability

- *Strength:* High | *Flexibility:* Low | *Durability:* Medium

### Printing Temperature (general)

- 190 °C – 230 °C}

### Bed Temperature (general)

- 0 °C – 60 °C

### Bed Adhesion

- Blue painters' tape/hairspray

### Additional Notes

Basic PLA can be brittle – avoid using it when making items that might be bent, twisted, or dropped repeatedly such as phone cases, high-wear toys, or tool handles.

You should also avoid using it with items which need to withstand higher temperatures, as PLA tends to deform around temperatures of 60 °C or higher. For all other applications, PLA makes for a good overall choice in filament.

# Polyvinyl alcohol (PVA)

**Notable Properties**

- Non-toxic and environmentally friendly, can easily be dissolved in water

**Typical Usage**

- Adhesives, thickeners, packaging film, various hygiene and incontinence products, children's play slime

**3D Print Application**

PVA is used for dissolvable supports for printers with dual extrusion 3D printers.

**Pros**

- Non-toxic and biodegradable
- Soluble in water
- Low flexibility and safe for food

**Cons**

- Can be difficult to use, as it absorbs moisture (very hygroscopic)
- Difficult to store
- Costly compared to other filaments

**Strength/Flexibility/Durability**

- Material not used for final print

**Printing Temperature (general)**

- 180 °C – 230 °C

**Bed Temperature (general)**

- ~45 °C

**Bed Adhesion**

- Blue painters' tape

**Additional Notes**

PVA a great support material when paired with another 3D printer filament in a dual extrusion 3D printer. The advantage of using PVA over HIPS is that it can be printed with more than just ABS, as it just uses water to dissolve.

The trade-off is the 3D printer filament that is slightly more difficult to handle. One must also be careful when storing it, as even the moisture in the atmosphere can damage the filament prior to printing. Dry boxes and silica pouches are a must if you plan to keep a spool of PVA usable in the long run.

# Polyethylene Terephthalate Glycol (PETG)

## Notable Properties

- FDA approved for food containers and tools used for food consumption, barely warps, no odors or fumes when printed

## Typical Usage

- Phone accessories, mechanical parts, water bottles, jewelry, props

## 3D Print Application

PETG is often used in phone/electronics cases and mechanical parts that require flexibility and impact resistance.

## Pros

- Food safe
- Easy material for 3D printing
- Flexible and mid-way between ABS and PLA
- Handles a wide temperature range without any issues
- Hard and shockproof

## Cons

- Absorbs moisture from the air. PETG must be stored properly.
- Requires higher temperature to print. Some printers cannot reach the required temperatures.

## Strength/Flexibility/Durability

- *Strength:* High | *Flexibility:* Medium | *Durability:* High

## Printing Temperature (general)

- 230 °C – 255 °C

## Bed Temperature (general)

- 55 °C – 70 °C

## Bed Adhesion

- Blue painters' tape

## Additional Notes

PETG is hygroscopic, which means it absorbs moisture from the air. As this has a negative effect on printing, make sure to store the 3D printer filament in a cool, dry place.

PETG is sticky during printing; it makes this 3D printer filament a poor choice for support structures but good for layer adhesion. (Just be careful with the print bed!)

Though not brittle, PETG scratches more easily than ABS.

## High Impact Polystyrene (HIPS)

### Notable Properties
- Biodegradable, 3D support material, low-cost

### Typical Usage
- Costume accessories, models, miniature figurines, general prototyping

### 3D Print Application
HIPS is popular as a support material in dual extrusion 3D printers to provide structural support to a complex object.

### Pros
- Similar to but typically prints better than ABS and slightly less likely to warp
- Very rigid

### Cons
- Has curling and adhesion issues similar to ABS but slightly less so
- Working would be tricky if you have no heated bed

### Strength/Flexibility/Durability
- *Strength:* High | *Flexibility:* Low | *Durability:* Medium

### Printing Temperature (general)
- 220 °C – 230 °C

### Bed Temperature (general)
- 50 °C – 60 °C

### Bed Adhesion
- Kapton Tape/hairspray

### Additional Notes
Used primarily for dual extrusion printing as removable support material for ABS. Immersing the finished print in limonene will strip away the HIPS, leaving your final product behind.

Unfortunately, using HIPS as a support material limits you to printing your actual part from ABS. Other 3D printer filament materials will be damaged by the limonene. Handily, HIPS and ABS print well together, being of similar strength, stiffness, and requiring a similar print temperature.

In fact, despite its primary use as a support material, HIPS is a decent 3D printer filament in its own right. It is very rigid, warps less than ABS, and can easily be glued, sanded, and painted.

## Nylon

### Notable Properties
- Strong, wear-resistant, lightweight, durable, flexible

### Typical Usage
- Machine parts, structural parts, mechanical components, gears, bearings, tools, various small consumer products

### 3D Print Application
Nylon is mainly used for its strength. It is also much less brittle than other plastics.

### Pros
- High strength, durability, and flexibility
- Less brittle than PLA
- Can be re-melted and used again without losing bonding properties
- Can be dyed

### Cons
- Very high melting temperature of at least 240 °C
- Hot-ends contain materials like PEEK and PTFE.
- When heated, can break down and emit toxic fumes.
- Hygroscopic

### Strength/Flexibility/Durability
- *Strength:* High | *Flexibility:* High | *Durability:* High

### Printing Temperature (general)
- 210 °C – 250 °C

### Bed Temperature (general)
- 60 °C – 80 °C

### Bed Adhesion
- PVA-based glue

### Additional Notes
Another unique characteristic of this 3D printer filament is that you can dye it either before or after the printing process. The negative side is that nylon, like PETG, is hygroscopic, meaning it absorbs moisture, so remember to store it in a cool, dry place to ensure better quality prints.


# › 6.4 – Are there hybrid materials?

There are many different types of filament, including hybrids. For example, PLA is a binder and a common medium for additives, making it ideal for adding other materials into the mix to create new types of filament with altered properties. A carbon fiber filament is carbon fiber mixed in with PLA, giving the material more strength than normal PLA.

Let's take a look at some hybrid materials:

## Carbon Fiber PLA

### Notable Properties
- Highly durable, low warpage, good layer adhesion

### Typical Usage
- Tools, hobbyist propellers, mechanical parts, protective casings, and various high durability applications

### 3D Print Application
This 3D printer filament is ideal for mechanical parts, protective casings, shells, and high durability applications.

### Pros
- Stronger than normal PLA
- PLA variant (PLA Pros)
- Prints well
- Very durable
- Does not require a heated bed
- Very little shrinkage and warping during cooling

## Cons

- Made of abrasive material that increases the wear and tear on printer nozzle
- Printer nozzle should be made or coated with a harder material. Common Brass nozzles will be damages quickly

### Strength/Flexibility/Durability

- *Strength:* High | *Flexibility:* Low | *Durability:* High

### Printing Temperature (general)

- 195 °C – 220 °C

### Bed Temperature (general)

- 0 °C – 60 °C

### Bed Adhesion

- Blue painters' tape

### Additional Notes

The tradeoff is the increased wear and tear on your printer's nozzle, especially if it is made of a soft metal like brass. As little as 500 grams of this exotic 3D printer filament will noticeably increase the diameter of a brass nozzle, so unless you enjoy frequently replacing your nozzle, consider using one made of (or coated with) a harder material.

# Metal PLA

### Notable Properties

- Highly durable, little shrinkage during cooling

### Typical Usage

- Jewelry, replicas, statues, home hardware

### 3D Print Application

Metal PLA produces metal-like products. It works great for jewelry, statues, home hardware, and artifact replicas. Metal PLA does not have the strength of metal, only the appearance.

### Pros

- PLA variant
- Highly durable
- Very little shrinkage during cooling

## Cons

- Requires fine-tuning of nozzle temperature, flow rate, and heavy post-processing
- Does not have the strength of metal.

### Strength/Flexibility/Durability

*Strength:* Variable | *Flexibility:* Low | *Durability:* High

### Printing Temperature (general)

- 195°C – 220 °C

### Bed Temperature (general)

- 50 °C

### Bed Adhesion

- Blue painters' tape

### Additional Notes

Bronze, brass, copper, aluminum, and stainless steel are just a few varieties of metal 3D printer filament which are commercially available. If there's a specific look you're interested in, don't be afraid to polish, weather, or patina your metal items after printing.

You may need to replace your nozzle a little sooner as a result of printing with metal. The grains are somewhat abrasive, resulting in increased nozzle wear.

The most common 3D printer filament blends tend to be around 50 percent metal powder and 50 percent PLA or ABS, but blends also exist that are up to 85 percent metal.

## Magnetic Iron PLA

### Notable Properties

- High durability, magnetic properties

### Typical Usage

- Fridge magnets, actuators, sensors, various other DIY projects

### 3D Print Application

Best used for its magnet, qualities and educational and DIY projects.

### Pros

- PLA variant
- Highly durable
- Heavy product

### Cons

- Requires fine-tuning of nozzle temperature, flow rate, and post-processing
- Due to density, 1 kg of filament is about half the length
- Can warp if do you quick cooling of parts
- Needs a heated bed
- Material is expensive

### Strength/Flexibility/Durability

- *Strength:* High | *Flexibility:* Low | *Durability:* Medium

### Printing Temperature (general)

- 185 °C

### Bed Temperature (general)

- 20 °C – 55 °C

### Bed Adhesion

- Blue painters' tape

### Additional Notes

Use this type of 3D printer filament whenever you want your prints to stick to something magnetic. Ornaments (especially for the refrigerator) are the most obvious example, but why not incorporate some magnets into toys or tools?

## Conductive PLA

### Notable Properties

- Very low warping, not soluble, prints low-voltage electronic circuits

### Typical Usage

- Low-voltage circuitry applications

### 3D Print Application

This 3D printer filament is ideal for LEDs, sensors, circuits and low-voltage Arduino projects.

### Pros

- PLA variant
- Allows low-voltage electronic circuits
- Does not require a heated bed

**Cons**

- Not durable and not very flexible
- Material may break under repeated bending
- Shrinks during cooling
- Conductive PLA 3D filament printer is expensive

**Strength/Flexibility/Durability**

- *Strength:* High | *Flexibility:* Low | *Durability:* Medium

**Printing Temperature (general)**

- 215 °C – 230 °C

**Bed Temperature (general)**

- 0 °C – 60 °C

**Bed Adhesion**

- Blue painters' tape

**Additional Notes**

Even though this 3D printer filament type only supports low-voltage circuitry, the sky's the limit with customized electronics projects. If you're experimenting, try coupling a circuit board with LEDs, sensors, or even a Raspberry Pi! If you're looking for something a little more specific, popular ideas include gaming controllers, digital keyboards, and trackpads.

## Glow-in-the-Dark PLA

**Notable Properties**

- Minimal warping, glow-in-the-dark properties, durable, low shrinkage during cooling

**Typical Usage**

- Children's toys, novelty items, wearables, phone accessories

**3D Print Application**

Glow-in-the-dark light switches, light-shades, toys, Halloween decorations

**Pros**

- PLA variant
- Durable, not soluble.
- Low shrinkage during cooling
- Does not require heated bed
- Similar to the standard PLA, printing is easy

## Cons

- All normal PLA cons

## Strength/Flexibility/Durability

- *Strength:* High | *Flexibility:* Low | *Durability:* Medium

## Printing Temperature (general)

- 185 °C – 205 °C

## Bed Temperature (general)

- 0 °C – 60 °C

## Bed Adhesion

- Blue painters' tape

## Additional Notes

This works by using phosphorescent materials mixed in with the PLA or ABS base. Thanks to these added materials, a glow-in-the-dark 3D printer filament is able to absorb and later emit photons, which are tiny particles of light. Your prints will only glow after being exposed to light.

For best results, consider printing with thick walls and little infill. The thicker your walls are, the stronger the glow.

# Wood PLA

## Notable Properties

- Versatility, real wood scent, durability, contain real wood fibers, stainable

## Typical Usage

- Sculptures, boxes, figurines, props

## 3D print Application

Wood filaments are often used in decors or materials that you want to achieve with a wooden look. And you can even stain the result for a wider range of looks.

## Pros

- Contains real wood fibers
- Produces different shades of brown and wooden-like surfaces
- The higher the temperature, the darker the brown shade becomes
- Changing the printing temperature simulates woodgrain
- You can decorate and post-process, like cut, grind, and paint

### Cons

- Softer and weaker compared to PLA
- Reduced flexibility and tensile length
- It can break easily

### Strength/Flexibility/Durability

- *Strength:* High | *Flexibility:* Low | *Durability:* Medium

### Printing Temperature (general)

- 195 °C – 220 °C

### Bed Temperature (general)

- 0 °C – 60 °C

### Bed Adhesion

- Blue painters' tape

### Additional Notes

There are many wood varieties, such pine, birch, cedar, ebony, and willow, but the range also extends to less common types like bamboo, cherry, coconut, cork, and olive.

Be careful with the temperature at which you print wood, as too much heat can result in an almost burnt or caramelized appearance. On the other hand, the base appearance of your wooden creations can be greatly improved with a little post-print processing! See **Figure 6.3**.

FIGURE 6.3 – Wood filament print. Jonathan Torta.

# › 6.5 – What specialty materials can I print with my FDM printer?

You might select some materials because their properties are completely different than common filament types. For example, some projects might need to be flexible – like tires, clothing, and mechanical joints. Others might need a special aesthetic look, like sandstone for architectural, museum, or landscape displays.

Next are some materials with different formulations:

## Flexible – Thermoplastic Elastomer (TPE)

### Notable Properties

- Flexible 3D printing materials, excellent abrasion resistance, smooth feeding properties, durability

### Typical Usage

- Toys, novelty items, wearable, phone accessories, automotive parts, household appliances, medical supplies, cosplay

### 3D Print Application

Ideal for anything that needs to be almost indestructible and flexible.

### Pros

- High elasticity and excellent abrasion resistance
- Consistent diameter, smooth feeding properties
- Sticks easily to build platform, bonds between layers for high-quality objects
- Durable, low shrinkage during cooling
- Does not require a heated bed

### Cons

- Can be difficult to print with and requires fine-tuning of nozzle temperature and flow rate
- Can get stuck in extruder

### Strength/Flexibility/Durability

- *Strength:* Medium | *Flexibility:* Very High | *Durability:* Very High

### Printing Temperature (general)

- 210 °C – 235 °C

### Bed Temperature (general)

- 20 °C – 55 °C

### Bed Adhesion

- Blue painters' tape

### Additional Notes

Thermoplastic polyurethane (TPU) is a variety of TPE and is itself a popular 3D printer filament. Compared to generic TPE, TPU is slightly more rigid, making it easier to print. It's also a little more durable and can better retain its elasticity in the cold.

Thermoplastic polyester (TPC) is another variety of TPE, though not as commonly used as TPU. Similar in most respects to TPE, TPC's main advantage is its higher resistance to chemical and UV exposure, as well to heat (up to 150°C).

See **Figure 6.4** for a sample of a flexible print.

FIGURE 6.4 – An example of a flexible print. Jonathan Torta.

## Sandstone

### Notable Properties

- Not plastic feeling. Printed objects can be colored and easily ground
- Sticks well on print bed
- No heated bed necessary and near zero warp

**Typical Usage**

- Architecture models, landscapes, table top gaming, sculpture

**3D Print Application**

This 3D printer is ideal for architectural, museum, or landscape display.

**Pros**

- Unique sandstone finish and appearance
- Does not require a heated bed
- Does not shrink or warp during cooling

**Cons**

- Not durable and less flexible
- Brittle and prone to snapping and breaking
- Material is not food safe

**Strength/Flexibility/Durability**

*Strength:* low | *Flexibility:* low | *Durability:* low

**Printing Temperature (general)**

- 165 °C – 210 °C

**Bed Temperature (general)**

- 0 °C – 60 °C

**Bed Adhesion**

- Blue painters' tape

**Additional Notes**

Can be brittle and small details can easily break off.

## Amphora

**Notable Properties**

- High strength and very high toughness
- FDA food contact compliance
- Odor neutral processing
- Styrene free formulation

**Typical Usage**

- Mechanical parts, food containers, cups, utensils, and bottles

### 3D Print Application

Amphora is stiff, lightweight and impact resistance, making it ideal for mechanical parts or containers that will hold food.

### Pros

- Produces little to no odor during printing
- Very strong
- Has a higher melting point than PLA
- Has better layer adhesion for improved surface finish
- Performs better when bridging gaps
- Has cleaner overhangs and has little to no warps
- Material is also US FDA-approved for food contact

### Cons

- Not as easy to print with
- Requires extensive fine-tuning of bed and nozzle temperatures

### Strength/Flexibility/Durability

*Strength:* High | *Flexibility:* medium when thin | *Durability:* High

### Printing Temperature (general)

- 220 °C – 250 °C

### Bed Temperature (general)

- 60 °C – 70 °C

### Bed Adhesion

- Blue painters' tape

### Additional Notes

A new polymer developed by Eastman for use in personal 3D printers. The material has extremely good safety qualities.

## Polyethylene coTrimethylene Terephthalate (PETT)

### Notable Properties

- Colorless, water clear, FDA approved, recyclable, strong, and flexible

### Typical Usage

- Food containers, cups, utensils and bottles

### 3D Print Application

PETT is an FDA approved polymer, making it safe for direct food contact. Most PETT applications include food containers like cups and utensils. Soda pop bottles are made of this material.

### Pros

- Strength, flexible, and biocompatibility. Neither brittle nor prone to warp.
- Does not shrink, can be printed on glass without any glues
- Does not absorb water or moisture from the air, does not degrade in water
- FDA approved and impressive in bridging
- Prices for PETT are also coming down, with some as cheap as ABS

### Cons

- Not easy to use
- Requires fine-tuning of bed and nozzle temperature

### Strength/Flexibility/Durability

- *Strength:* High | *Flexibility:* Medium | *Durability:* High

### Printing Temperature (general)

- 210 °C – 230 °C

### Bed Temperature (general)

- 45 °C

### Bed Adhesion

- Blue painters' tape

## Additional Specialty Materials

There are some materials that are not as common but are still used to enhance certain projects. We list a few as examples.

## Acrylonitrile Styrene Acrylate (ASA)

In addition to being a 3D printer filament that is strong, rigid, and relatively easy to print with, ASA is also extremely resistant to chemical exposure, heat, and mostly importantly, changes in shape and color. Prints made of ABS have a tendency to denature and yellow if left outdoors. This is not the case with ASA.

Another minor benefit to using ASA over ABS is that it warps less during printing. Be careful with how you adjust your cooling fan; ASA can easily crack if things get a little too windy during printing.

## ACETAL (POM)

Acetal as a material sees common use as gears, bearings, camera focusing mechanisms, and zippers.

POM performs exceptionally well in these types of applications due to its strength, rigidity, resistance to wear, and most importantly its low coefficient of friction. It's thanks to this last property that POM makes such a great 3D printer filament.

For most of the types of 3D printer filament in this list, there is a significant gap between what is made in industry and what you can make at home with your 3D printer. For POM, this gap is somewhat smaller; the slippery nature of this material means prints can be nearly as functional as mass-produced parts.

Make sure to use a heated print bed when printing with POM 3D printer filament as the first layer doesn't always want to stick.

## PMMA (Acrylic)

3D printing with PMMA filament can be a little difficult. To prevent warping and to maximize clarity, extrusion must be consistent, which requires a high nozzle temperature. It is also helpful to enclose the print chamber in order to better regulate cooling.

## Flexible polyester (FPE)

Two notable aspects of FPE include good layer-to-layer adhesion and a moderately high resistance to heat and a variety of chemical compounds. Given the wide range of FPE 3D printer filament that is available, perhaps the most useful way to differentiate between the wide ranges of FPE available is the Shore value (like 85A or 60D), where a higher number indicates less flexibility.

## SUMMARY

There are many things to consider when deciding which of the many materials should be used, such as: size, type, cost, availability, and quality. Some other things to consider are the materials' properties, usage, application, pros, cons, strengths, temperature, and bed adhesions. Also consider the wide variety of hybrid materials.

## APPLYING WHAT YOU'VE LEARNED

1. Continue making your own 3D dictionary by adding the definition (in your own words) of five words related to 3D printing in this chapter.

2. Describe 12 common materials used in 3D printers.

3. Discuss how to select materials for a 3D printer.

4. Pick five common basic filaments for 3D printers and describe them.

5. Choose four hybrid materials for 3D printers and write about their advantages and disadvantages.

6. If you have a 3D printer, describe what materials you would use for a project you will make or hope to make.

7. Decide on a project and explain what materials you would use and why.

## REFERENCES

[1] – https://all3dp.com/1/3d-printer-filament-types-3d-printing-3d-filament/

[2] – https://www.homedepot.com/s/3d%2520printer%2520filament?NCNI-5

[3] – https://www.amazon.com/s/ref=nb_sb_ss_i_4_4?url=search-alias%3Daps&field-keywords=3d+printer+filament&sprefix=3d+p%2Caps%2C219&crid=21FEUJ0XPTB33

[4] – https://edutechwiki.unige.ch/en/3D_printer_filament

# CHAPTER 7

By Dirk van der Made

# Applications and Slicing Settings

## OVERVIEW AND LEARNING OBJECTIVES

**In this chapter:**

- 7.1 – What types of applications are used for 3D printing?
- 7.2 – What software can I use to create a 3D model?
- 7.3 – What are some slicing applications?
- 7.4 – What are some common basic slicer settings?
- 7.5 – What are some common advanced slicer settings?
- 7.6 – Are there additional software applications used in 3D printing?

## › 7.1 – What types of applications are used for 3D printing?

There are many types of software programs or applications with a full range of features used for 3D model creation and slicing stages. In this chapter we will list a few programs for both stages, along with highlighting some of their features.

Because slicing (dividing the 3D model into layers for printing) is unique, we will review some basic common and advanced settings you may see in your slicer application's interface in Section 7.4 of this chapter. These slicer settings control the division of your 3D model into printable layers.

The software needed for 3D printing can be divided up into four stages (see **Figure 7.1**):

| 3D model creation | Slicing | Transferring | Monitoring |

FIGURE 7.1 – 3D printing application types.

- **3D model creation** – creates or fixes a 3D model and exports it into a usable file format
- **Slicing** – divides the 3D model into printable layers
- **Transferring** – sends your files to the printer
- **Monitoring** – oversees the printer during operation

## › 7.2 – What software can I use to create a 3D model?

3D modeling software ranges from beginner to advanced, free to commercial, and varies by operating system. Each program has a different interface, set of tools, and export settings. It is best to do your research to find the application that best fits into your skillset, budget, and workflow.

Some points to look for:

- **File formats** – Make sure the application opens and exports the formats you are using in your workflow.
- **Features** – Every application has a different set of tools. Make sure the application fits your needs.
- **Complexity** – Examine the learning curve of the application and see if there are learning tools like tutorials to help.
- **Operating system** – Verify that the application will run on your computer.
- **Overall Cost** – The application should fit your budget with the feature set you need.

Some applications are better suited to certain types of projects. For example, CAD programs are well suited for precise measurements and ideal for mechanical objects or parts while animation or sculpting programs might be better suited for artistic projects.

**MAKER'S NOTE**

In **Figure 7.2** we see an example of the interface for the free software MeshLab loaded with a 3D model.

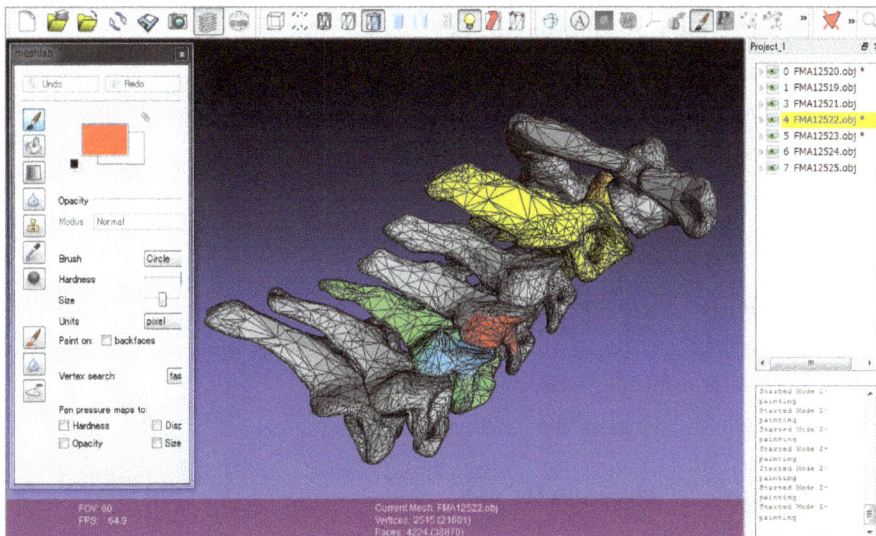

FIGURE 7.2 – Painting colors on cervical vertebrae polygon data which are included in free human body data, BodyParts3D. Software used for painting is free software MeshLab. Polygon data is BodyParts3D.

# Software for creating and editing objects
## Beginner

- **Tinkercad® by Autodesk, Inc.**
  - › *Type:* creation
  - › *Operating system:* web browser
  - › *Cost:* free
  - › *Readable formats:* .123dx, .3ds, .c4d, .mb, .obj, .svg, .stl

- **LibreCAD by LibreCAD**
  - › *Type:* CAD, creation
  - › *Operating system:* Windows, macOS, Linux
  - › *Cost:* free
  - › *Readable formats:* .dxf, .dwg

- **3D Slash by 3D Slash**
  - › *Type:* logo maker
  - › *Operating system:* Windows, macOS, Linux, Raspberry Pi, web browser
  - › *Cost:* free standard edition, commercial full edition
  - › *Readable formats:* .3dslash, .obj, .stl

- **SculptGL by Stephane Ginier**
  - › *Type:* sculptor
  - › *Operating system:* web browser
  - › *Cost:* free
  - › *Readable formats:* .obj, .ply, .sgl, .stl

- **SelfCAD by SelfCAD**
  - › *Type:* CAD, creation
  - › *Operating system:* web browser
  - › *Cost:* commercial, free trial
  - › *Readable formats:* .dae, .mtl, .ply, .stl, .svg

- **Sculptris by Pixologic™**
  - › *Type:* sculptor
  - › *Operating system:* Windows

> *Cost:* free

> *Readable format:* .obj

- **MeshMagic 3D Modeling Software by NCH Software®**
  > *Type:* editor, tool
  > *Operating system:* Windows
  > *Cost:* free
  > *Readable formats:* .3dp, .3ds, .3mf, .obj, .ply, .stl

- **Leopoly by Leopoly LTD**
  > *Type:* creation
  > *Operating system:* web browser, virtual reality (VR)
  > *Cost:* free standard edition, commercial full edition
  > *Readable format:* .stl

- **3D Builder by Microsoft® Corporation**
  > *Type:* creation, editor, tool
  > *Operating system:* Windows
  > *Cost:* free
  > *Readable formats:* .3mf, .obj, .ply, .stl

## Intermediate

- **Meshmixer by Autodesk, Inc.**
  > *Type:* creation, editor, tool
  > *Operating system:* Windows, macOS, Linux
  > *Cost:* free
  > *Readable formats:* .amf, .mix, .obj, .off, .stl

- **MeshLab by MeshLab**
  > *Type:* creation, editor, tool
  > *Operating system:* Windows, macOS, Linux
  > *Cost:* free
  > *Readable formats:* .3ds, .collada, .obj, .off, .ply, .stl, .u3d, .vrml , .x3d,

- **SketchUp by Trimble Inc.**
  > *Type:* creation

- › *Operating system:* Windows, macOS
- › *Cost:* free standard edition, commercial full edition
- › *Readable formats:* .3ds, .dae, .def, .dem, .dwg, .dxf, .ifc, .kmz, .stl

- ■ **FreeCAD by FreeCAD**
  - › *Type:* CAD, creation
  - › *Operating system:* Windows, macOS, Linux
  - › *Cost:* free
  - › *Readable formats:* .dae, .dxf, .fcstd, .ifc, .iges, .nastran, .obj, .off, .step, .stl, .svg

- ■ **OpenSCAD by OpenSCAD**
  - › *Type:* CAD, creation
  - › *Operating system:* Windows, macOS, Linux
  - › *Cost:* free
  - › *Readable formats:* .dxf, .off, .stl

- ■ **MakeHuman™ by MakeHuman**
  - › *Type:* animation, creation, human modeler
  - › *Operating system:* Windows, macOS, Linux
  - › *Cost:* free
  - › *Readable formats:* .dae, .fbx, .obj, .stl

- ■ **nanoCAD by Nanosoft**
  - › *Type:* CAD, creation
  - › *Operating system:* Windows
  - › *Cost:* free standard edition, commercial full edition
  - › *Readable formats:* .3dm, .dae, .dfx, .dwg, .dwt, .iges, .igs, .pdf, .sat, .sldprt, .ssm_bin, .step, .stl, .x_b, .x_t, .xxm_txt

- ■ **DesignSpark by RS Components Limited**
  - › *Type:* CAD, creation
  - › *Operating system:* Windows
  - › *Cost:* free standard edition, commercial full edition
  - › *Readable formats:* .dxf, .ecad, .emn, .idb, .idf, .iges, .obj, .rsdoc, .skp, .step, .stl

- **Clara.io by Exocortex Technologies, Inc.**
  › *Type:* creation
  › *Operating system:* web browser
  › *Cost:* free standard edition, commercial full edition
  › *Readable formats:* .3dm, .3ds, .cd, .dae, .dgn, .dwg, .emf, .fbx, .gf, .gdf, .gts, .igs, .kmz, .lwo, .rws, .obj, .off, .ply, .pm, .sat, .scn, .skp, .slc, .sldprt, .stp, .stl, .x3dv, .xaml, .vda, .vrml, .x_t, .x, .xgl, .zpr

- **Moment of Inspiration (MoI) by Triple Squid Software Design**
  › *Type:* CAD lite, creation
  › *Operating system:* Windows, macOS
  › *Cost:* commercial
  › *Readable formats:* .3ds, .3dm, .dxf, .fbx, .igs, .lwo, .obj, .skp, .stl, .stp, .sat

## Professional

- **AutoCAD® by Autodesk, Inc.**
  › *Type:* CAD, creation
  › *Operating system:* Windows, macOS
  › *Cost:* commercial
  › *Readable formats:* .dwg, .dxf, .pdf

- **Blender® by Blender Foundation**
  › *Type:* animation, creation, modeler
  › *Operating system:* Windows, macOS, and Linux
  › *Cost:* free
  › *Readable formats:* .3ds, .dae, .dxf, .fbx, .lwo, .obj, .ply, .stl, .svg, .vrml, .vrml97, .x, .x3d

- **Lightwave® by NewTek Inc.**
  › *Type:* animation, creation, modeler
  › *Operating system:* Windows, macOS
  › *Cost:* commercial, free trial
  › *Readable formats:* .3ds, .dae, .dxf, .fbx, .obj, .ply, .stl

- **Maya® by Autodesk, Inc.**
  › *Type:* animation, creation, modeler

> › *Operating system:* Windows, macOS and Linux
> › *Cost:* commercial, free trial
> › *Readable formats:* .abc, .abc, .ai, .anim, .apf, .cat, .cat, .catproduct, .csb, .dae, .dfx, .dwg, .editMA, .editMB, .eps, .fbx, .flt, .iam, .iges, .iges, .ipt, .it, .itp, .iv, .jt, .ma, .mb, .mel, .model, .mov, .nx, .obj, .prt, .prt, .sat, .sldprt, .step, .stl, .stl, .stp, .stp, .wire

- **Cinema 4D by MAXON Computer GmbH**
  > › *Type:* animation, creation, modeler
  > › *Operating system:* Windows, macOS
  > › *Cost:* commercial
  > › *Readable formats:* .3ds, .dae, .dem, .dwg, .dxf, .fbx, .iges, .lwf, .obj, .rib, .skp, .stl, .wrl, .x

- **3ds Max® by Autodesk, Inc.**
  > › *Type:* animation, creation, modeler
  > › *Operating system:* Windows
  > › *Cost:* commercial, educational
  > › *Readable formats:* .3ds, .abc, .ai, .ase, .asm, .catpart, .catproduct, .dem, .dwf, .dwg, .dxf, .flt, .iges, .ipt, .jt, .nx, .obj, .prj, .prt, .rvt, .sat, .skp, .sldasm, .sldprt, .stl, .stp, .vrml, .w3d, xml

- **ZBrush by Pixologic, Inc.**
  > › *Type:* sculptor
  > › *Operating system:* Windows, macOS
  > › *Cost:* commercial, educational
  > › *Readable formats:* .dxf, .goz, .ma, .obj, .stl, .vrml, .x3d

- **Modo® by The Foundry Visionmongers Ltd.**
  > › *Type:* animation, creation, modeler
  > › *Operating system:* Windows, macOS, Linux
  > › *Cost:* commercial
  > › *Readable formats:* .3dm, .abc, .dae, .dxf, .fbx, .geo, .lwo, .obj, .pdb, .stl,.x3d

- **Onshape by Onshape Inc.**
  > › *Type:* CAD, creation

> › *Operating system:* Windows, macOS, Linux, iOS, Android
> › *Cost:* commercial
> › *Readable formats:* .3dm, .dae, .dfx, .dwg, .dwt, .iges, .igs, .pdf, .sat, .sldprt, .ssm_bin, .step, .stl, .x_b, .x_t, .xxm_txt

- **Poser® by Smith Micro Software, Inc.**
  > › *Type:* animation, creation, human modeler
  > › *Operating system:* Windows, macOS
  > › *Cost:* commercial
  > › *Readable formats:* .cr2, .obj, .pz2

- **Rhinoceros (Rhino) 3D by Robert McNeel & Associates**
  > › *Type:* creation
  > › *Operating system:* Windows, macOS
  > › *Cost:* commercial, educational
  > › *Readable formats:* .3dm, .3ds, .cd, .dae, .dgn, .dwg, .emf, .fbx, .gf, .gdf, .gts, .igs, .kmz, .lwo, .rws, .obj, .off, .ply, .pm, .sat, .scn, .skp, .slc, .sldprt, .stp, .stl, .x3dv, .xaml, .vda, .vrml, .x_t, .x, .xgl, .zpr

- **Mudbox® by Autodesk, Inc.**
  > › *Type:* sculptor
  > › *Operating system:* Windows, macOS
  > › *Cost:* commercial
  > › *Readable formats:* .fbx, .mud, .obj

## Industrial

- **SOLIDWORKS by Dassault Systèmes**
  > › *Type:* CAD, creation
  > › *Operating system:* Windows
  > › *Cost:* commercial, educational
  > › *Readable formats:* .3dxml, .3dm, .3ds, .3mf, .amf, .dwg, .dxf, .idf, .ifc, .obj, .pdf, .sldprt, .stp, .stl, .vrml

- **Inventor® by Autodesk, Inc.**
  > › *Type:* CAD, creation
  > › *Operating system:* Windows, macOS

> › *Cost:* commercial
> › *Readable formats:* .3dm, .igs, .ipt, .nx, .obj, .prt, .rvt, .sldprt, .stl, .stp, .x_b, .xgl

- **Fusion 360™ by Autodesk, Inc.**
  - › *Type:* CAD, creation
  - › *Operating system:* Windows, macOS
  - › *Cost:* free standard edition, commercial full edition, educational
  - › *Readable formats:* .catpart, .dwg, .dxf, .f3d, .igs, .obj, .pdf, .sat, .sldprt, .stp

- **CATIA by Technia Transcat**
  - › *Type:* CAD, creation
  - › *Operating system:* Windows
  - › *Cost:* commercial, educational
  - › *Readable formats:* .3dxml, .catpart, .igs, .pdf, .stp, .stl, .vrml

## 3D Scanner

Most 3D scanners come with software that is made for the scanner. If possible, it is best to use the application that comes with the scanner. However, there are other types of 3D scanning applications that could be used.

- **Autodesk 123D Catch** by Autodesk, Inc.
- **itSeez3D©** by Itseez3D, Inc.
- **Qlone** by EyeCue Vision Technologies LTD
- **Einscan©** by Shining3D®
- **Matter and Form 3D Scanner** by Matter and Form, Inc.

## Cell Phone Camera Apps

Although phones are not the best 3D scanners, there are model applications that you can download. These are changing every day. We recommend that you research the current available apps if you use your cell phone for scanning.

## Dedicated Object Creation

There are a few applications dedicated to specific object creation.[1]

- **CandleCaster by OpenJSCAD.org** – Creates candle holders

- **Cookiecaster by dreamforge, Inc.** – Creates cookie molds from a model or a photo
- **3D Racers by 3DRacers Ltd** – Creates racing cars
- **Terrain2STL by Thatcher Chamberlin** – Creates topographic maps; you can choose the areas on the website
- **Printshop MakerBot by MakerBot Industries, LLC** – Creates and print different models such as 3D printed jewelry bracelets or decorative objects
- **Image to Lithophane by Nested Cube** – Creates 3D from a photo that you download onto the platform; you can make a Lithophane (an engraved model) that you can only be seen if you place it in the light
- **Customizer by Thingiverse** – Creates parametric objects that people have designed

## › 7.3 – What are some slicing applications?

Slicing applications run the full spectrum of features and limitations; some may be rather basic with general settings, others may have more advanced settings, giving the maker full control. Most 3D printers come with their own proprietary software. (See **Figure 7.3.**)

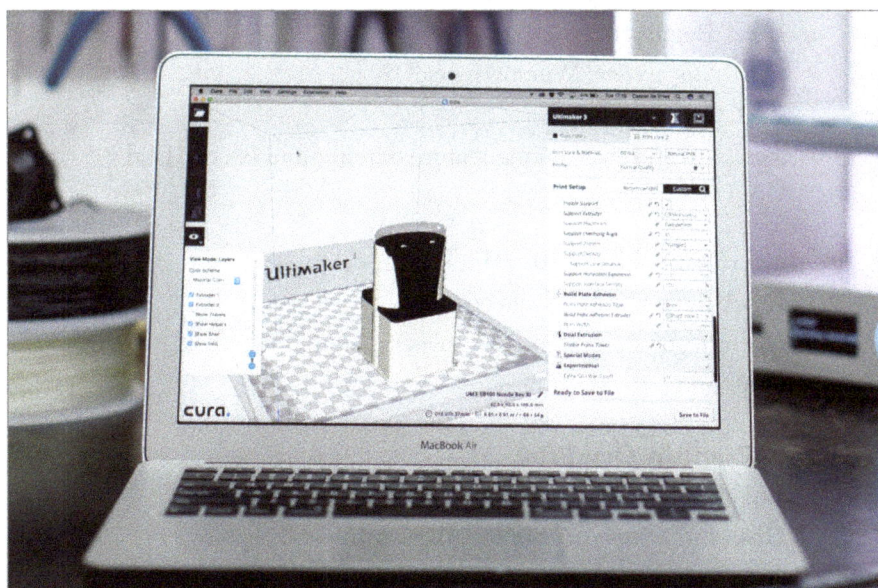

FIGURE 7.3 - Preparing a model for 3D printing using Ultimaker Cura software. Svitlana Lozova.

Whichever application you use, be sure to find one that works for your printer and has the settings you need for your basic or specialized projects.

Here is a list of slicing applications with additional notes on their difficulty level, operating system, and cost.

## Slicing software

- **Astroprint® by Astroprint**
  - › *Level:* Beginner to Advanced
  - › *Operating system:* Browser, Raspberry Pi, pcDuino
  - › *Cost:* free, commercial
  - › *Notes:* Raspberry Pi, pcDuino client, remote access

- **Ultimaker Cura by Ultimaker**
  - › *Level:* Beginner to Advanced
  - › *Operating system:* Windows, macOS, Linux
  - › *Cost:* open source slicer made for Ultimaker
  - › *Notes:* can be used by most printers, large number of features, many new updates

- **MakerBot Print by MakerBot Industries, LLC**
  - › *Level:* Beginners
  - › *Operating system:* Windows, macOS
  - › *Cost:* free
  - › *Notes:* priority, has unique feature of minimum fill, can load CAD files

- **MatterControl by MatterHackers Inc.**
  - › *Level:* Beginner to Advanced
  - › *Operating system:* Windows, macOS, Linux
  - › *Cost:* free

- **OctoPrint® by OctoPrint**
  - › **Level:** Intermediate to Advanced
  - › *Operating system:* Windows, macOS, Linux, Raspberry Pi
  - › *Cost:* free
  - › *Notes:* Raspberry Pi, remote access

- **Repetier by Hot-World GmbH & Co. KG**
  - › *Level:* Intermediate to Advanced
  - › *Operating system:* Windows, macOS, Linux
  - › *Cost:* free
  - › *Notes:* open source slicer software supports three different slicing engines (Slic3r, CuraEngine, and Skeinforge), visualizer, remote access

- **Simplify3D® by Simplify3D**
  - › *Level:* Beginners, Advanced
  - › *Operating system:* Windows, macOS
  - › *Cost:* commercial
  - › *Notes:* extremely powerful, can be used with just about any printer, large feature set

- **Slic3r by Slic3r**
  - › *Level:* Advanced
  - › *Operating system:* Windows, macOS, Linux
  - › *Cost:* free
  - › *Notes:* open source software, fast, flexible, and precise

- **3DPrinterOS by 3D Control Systems Ltd.**
  - › *Level:* Beginners to Advanced
  - › *Operating system:* web browser, Windows, macOS
  - › *Cost:* free, commercial
  - › *Notes:* cloud 3D printer management software

- **Craftware by Craftunique, LLC**
  - › *Level:* Beginners to Advanced
  - › *Operating system:* Windows, macOS
  - › *Cost:* free
  - › *Notes:* Good basic slicer with good visualizer

- **Flashprint by FlashForge Corporation**
  - › *Level:* Beginners to Advanced
  - › *Operating system:* Windows
  - › *Cost:* free

- **IceSL by INRIA / Sylvain Lefebvre**
  - › *Level:* Advanced
  - › *Operating system:* Windows, Linux
  - › *Cost:* free

- **KISSlicer by KISSlicer**
  - › *Level:* Beginners to Advanced
  - › *Operating system:* Windows, macOS, Linux, Raspberry Pie
  - › *Cost:* commercial
  - › *Notes:* Basic interface, does what it does well

- **Netfabb Standard by Autodesk Inc.**
  - › *Level:* Intermediate to Advanced
  - › *Operating system:* Windows
  - › *Cost:* commercial
  - › *Notes:* can identify issues with your STL files before they get to the slicing stage

- **ReplicatorG by ReplicatorG**
  - › *Level:* Intermediate to Advanced
  - › *Operating system:* Windows, macOS, Linux
  - › *Cost:* free
  - › *Notes:* open source slicer – a bit dated

- **SelfCAD by SelfCAD**
  - › *Level:* Beginner to Advanced
  - › *Operating system:* web browser
  - › *Cost:* commercial
  - › *Notes:* Web based slicer, highly compatible with object library, building and editing included

- **SliceCrafter by Sylvain Lefebvre / Inria**
  - › *Level:* Advanced
  - › *Operating system:* web browser
  - › *Cost:* free
  - › *Notes:* Web based slicer, highly compatible

- **Tinkerine Suite™ by Studios Ltd.**
  - › *Level:* Beginners
  - › *Operating system:* Windows, macOS
  - › *Cost:* free
  - › *Notes:* Basic interface, highly compatible

- **Z-Suite by Zortrax**
  - › *Level:* Beginners
  - › *Operating system:* Windows, macOS
  - › *Cost:* free
  - › *Notes:* priority, thin wall and object analysis and repair, editable support

# › 7.4 – What are some common basic slicer settings?

Most slicing applications have a set of five main basic settings. These slicer settings are the cornerstones of your 3D print.

1. Layer Height
2. Fill Density
3. Supports
4. Platform Adhesion – skirt, brim, raft
5. Shell Thickness

Often slicer applications have preset profiles that change most of the basic settings for you to best fit the needs of your print. These preset profiles can be generic, printer based, filament type, or even filament brand specific.

It is important to remember that these basic settings can be changed interdependently and they have little effect on each other. The slicer application takes care of the large set of sub- or advanced settings as needed.

However, most slicer applications have advanced settings. If you use and change the more advanced settings in the slicer application, they can be directly dependent on other settings, and must be changed carefully. Keep in mind their useable range and proportional/inversely proportional relations and interactions to other settings.

If you are just starting out, we recommend that you only adjust the basic settings before fine-tuning with the more advanced settings. Once you have more experience in how all the settings work in conjunction with each other, then proceed to work on one advanced setting at a time. For more information on advanced settings see Chapter 9.

Understanding the basic slicer settings is important; it provides a strong foundation to build upon. Because of this, in this section we will briefly describe the common basic settings for overall understanding and then follow up with a more detailed explanation to extend that understanding.

## Layer height
### Short Description

Layer height is the vertical resolution of your print and each individual layer that your printer must deposit. This setting specifies the height of each filament layer in your print. Think of slicing a food item like a carrot or apple – you have an object cut into slices that you can disassemble and reassemble. The thickness of the slices is the layer height. (See **Figure 7.4**.)

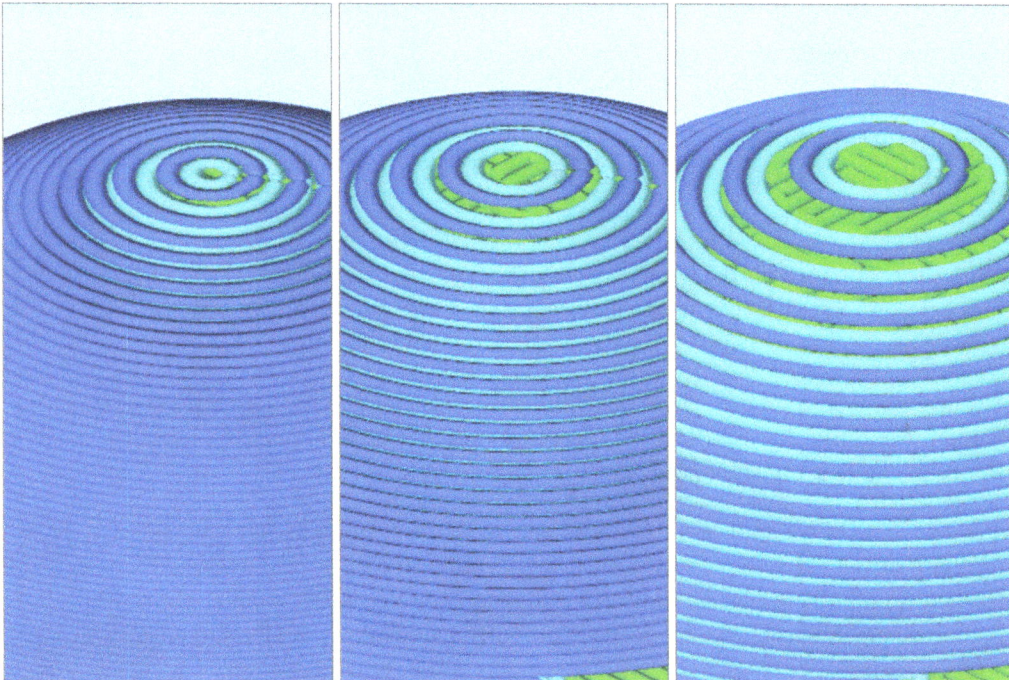

FIGURE 7.4 – Examples at three different layer heights: 0.1 mm, 0.2 mm, and 0.3 mm. Note how the steps at the top get more pronounced the thicker the layer. Jonathan Torta.

## Detailed Explanation

Setting the layer height properly is important, as the resolution can be limited by your printer and can affect other factors. Most printers have a good range of heights that can be used, but often just set a few defaults like Fine/High, Average/Medium and Coarse/Fast.

This actual layer height range is dictated by two factors: the nozzle diameter and the mechanical precision of your printer. A typical 0.4 mm nozzle can handle a usable layer height between 0.1 mm and 0.3 mm (though some can reach below 0.1 mm). Smaller and larger nozzles can handle different ranges. The lowest end of this range is dictated less by the nozzle than by the mechanical exactness of your printer and by the material itself. Some materials do not flow well in proximity to the bed or previous layer. On the high end, the layer height must be less than the diameter of the nozzle as the filament needs to be squashed into the lower layer in order to adhere. Too large a layer height in relation to your nozzle and you start losing layer cohesion. Your print will delaminate (divide into layers).

Given you have such a range of resolution choices, you may expect to always print at the highest resolution possible the printer can handle for the most detail, but time and reliability is a big factor. As the following example demonstrates, when the only modified variable is the layer height, the build time, the overall filament used, and the weight of the print are affected.

Common layer heights for a 0.4 mm nozzle. (See **Figures 7.5 through 7.7**.)

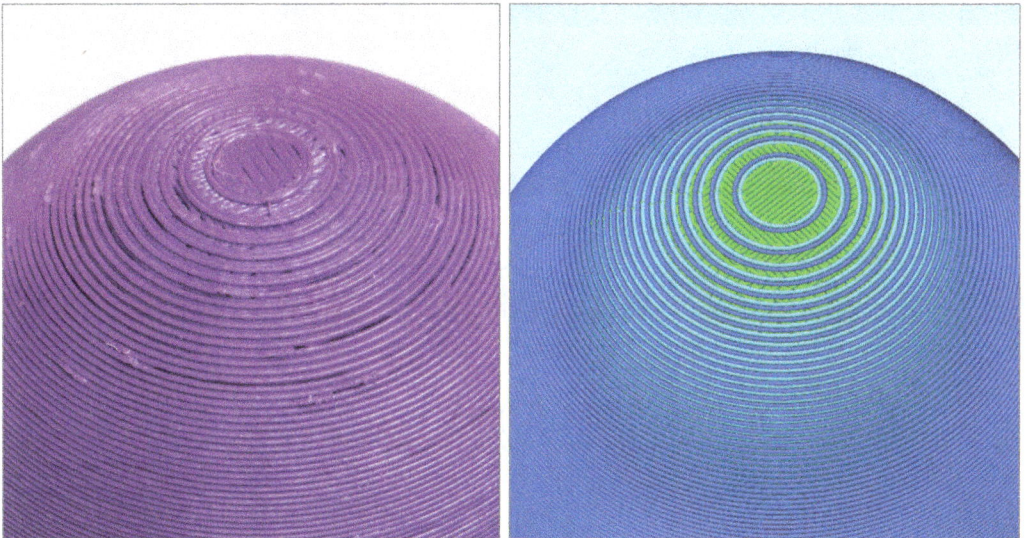

FIGURE 7.5 – 0.3 mm. Jonathan Torta.

## 0.3 mm

This is considered coarse. It prints fast and is the strongest. There are very obvious layers and it can take some effort to hide/remove them in post cleanup.

- **Build time:** 0 hours 25 minutes
- **Filament length:** 1985.6 mm
- **Plastic weight:** 5.92 g
- **Layers:** 69

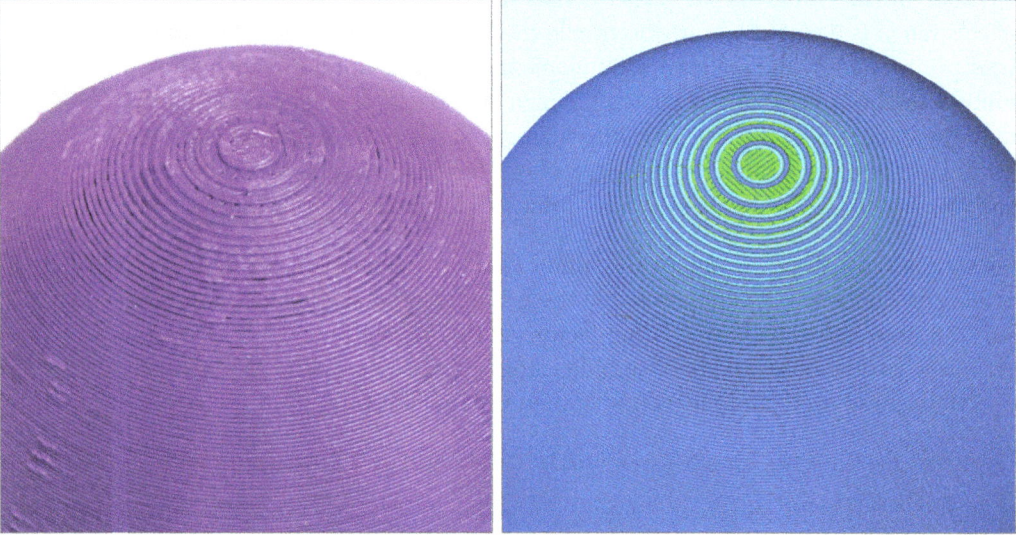

FIGURE 7.6 – 0.2 mm. Jonathan Torta.

## 0.2 mm

This is a great compromise of speed and quality. The layers are fine enough to look good or, if refining, take less effort to smooth but are also fast enough for day-to-day use.

Note that 0.2 mm is two times 0.1 mm but not half of 0.3 mm (the reason for the jump in the numbers between 0.2 mm and 0.1 mm layer heights).

- **Build time:** 0 hours 37 minutes
- **Filament length:** 1881.0 mm
- **Plastic weight:** 5.61 g
- **Layers:** 105

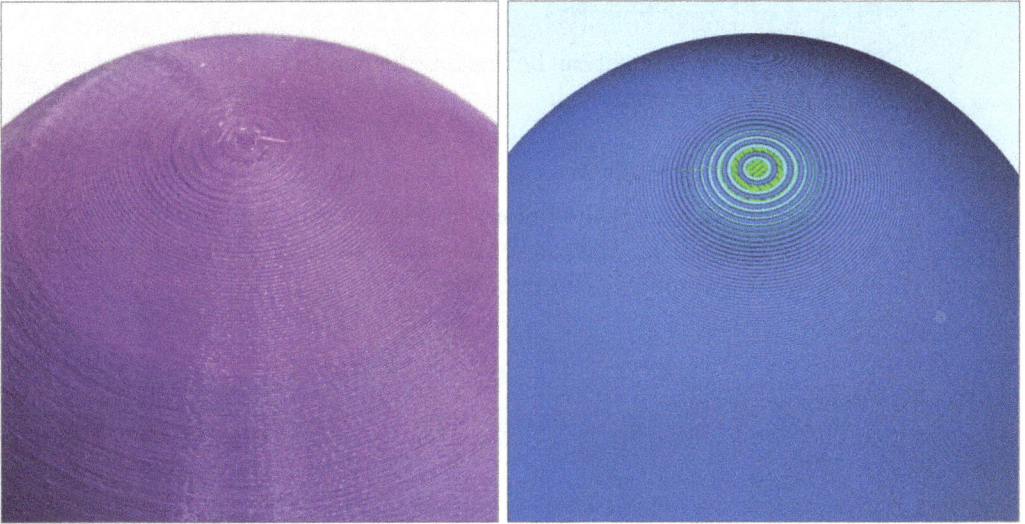

FIGURE 7.7 – .1 mm. Jonathan Torta.

## 0.1 mm

This is normally the finest detail average printers can print reliably; very smooth height resolution. This also can take substantially longer to print.

- **Build time:** 1 hour 11 minutes
- **Filament length:** 1793.0 mm
- **Plastic weight:** 5.35 g
- **Layers:** 209

Each layer height step decrease in turn increases the time required to print greatly. This is primarily due to just having to do more actions to do. I.e. moving from 0.2mm to 0.1mm should be 2x the time. This is not exact because of the way the printer prints some features. We will look more into this in Chapter 12.

The thinner the layer height, the better the detail (on the z-axis), with the cost that more layers take much more time to print. The thicker the layer height, the faster your print will be, but it will have less detail (most notably "stair-stepping" on curves and horizontal acute angles). Generally, the amount of material used decreases slightly with layer height, but this is not always the case and it can change depending on other factors.

Note that a 0.4 mm nozzle 3D printer does not extrude half the amount of filament that a 0.8 mm nozzle does – halving the nozzle diameter actually reduces the amount of filament extruded by 25%.

## Effects of Layer Height

The effects of layer height can be dramatic, depending greatly on the object and orientation you are printing in.

## Layers Making Curves

When printing at a lower resolution, the edges of your print will have a rougher, stepped edge to them. The thicker the layers become, the larger the steps between layers become (see **Figure 7.8**).

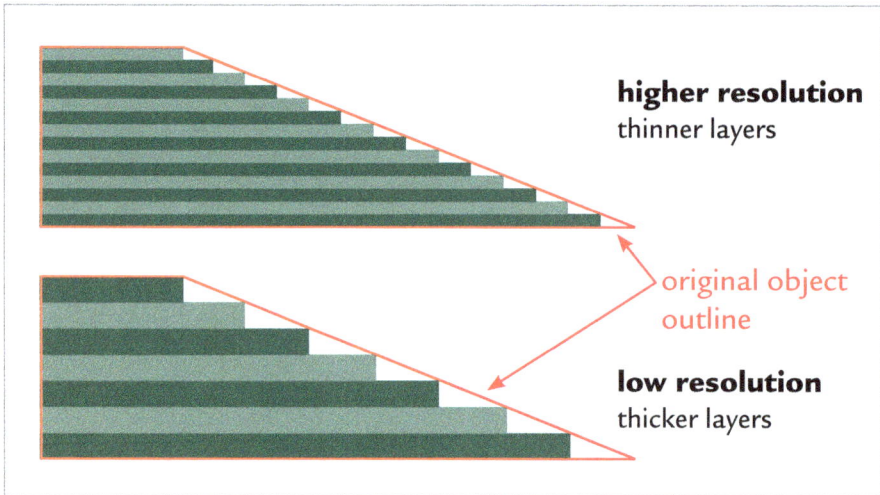

FIGURE 7.8 – Illustration showing high and low resolution.

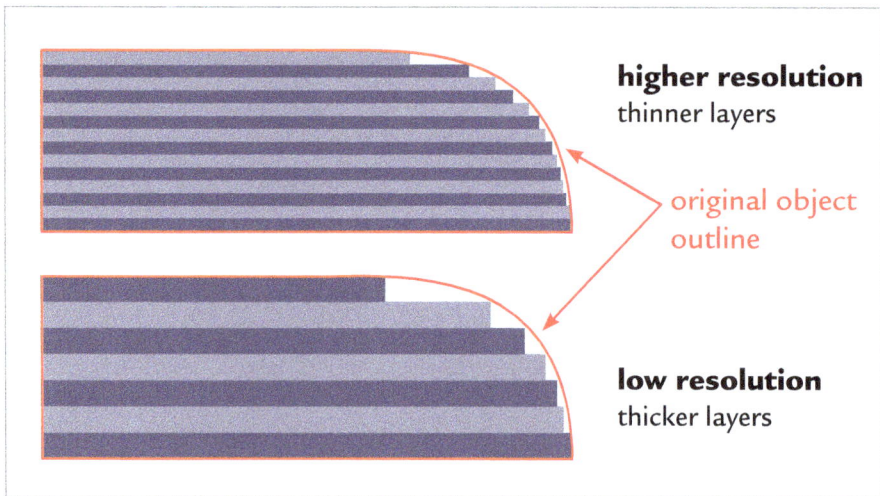

FIGURE 7.9 – Illustration showing stepped curves

Thinner layers produce a larger number of smaller steps, and these smooth out the edge of the curve. This is similar to high-and low-resolution images of circles. The lower the resolution of the image, the bigger the pixels and the more pixelated or jagged the edges of the circle appear (see **Figure 7.9**).

When possible, orient your prints so that curves are parallel with your layers. This allows the curve to be drawn with filament rather than be made by stacked layers.

> Very shallow curves in the X/Y plane tend to give a topographic map appearance and are not ideal. (See Figures 7.10–11.)

## Shallow/Acute Angles

The stepped issue with curves is even more exaggerated with low, horizontal shallow or acute angles. Very shallow slopes are defined by only a few layers and tend to be quite noticeable. These are best printed in vertical orientation if possible, making them effectively a very steep angle defined by many steps, printing much smoother. (See **Figures 7.10–12**.)

### 0.3 mm

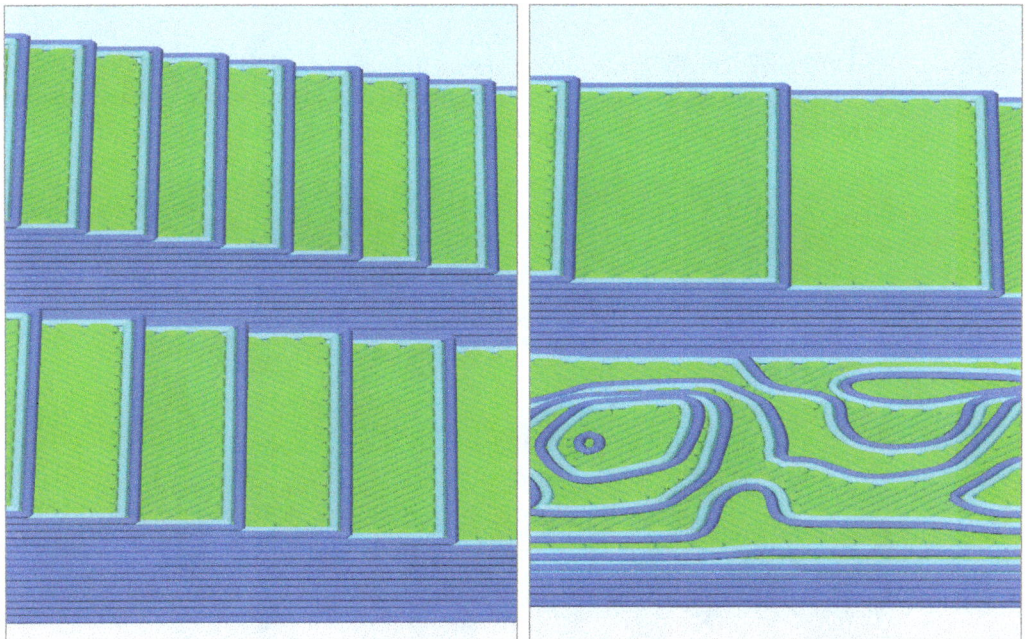

FIGURE 7.10 – High layer height/low resolution of 0.3 mm with angles 6 and 4 (left top and bottom) and 2 degrees top right. Also, a low curved space showing the topographic map effect (bottom right). Note the top layer is highlighted in green and the shell/perimeter in blue. Jonathan Torta.

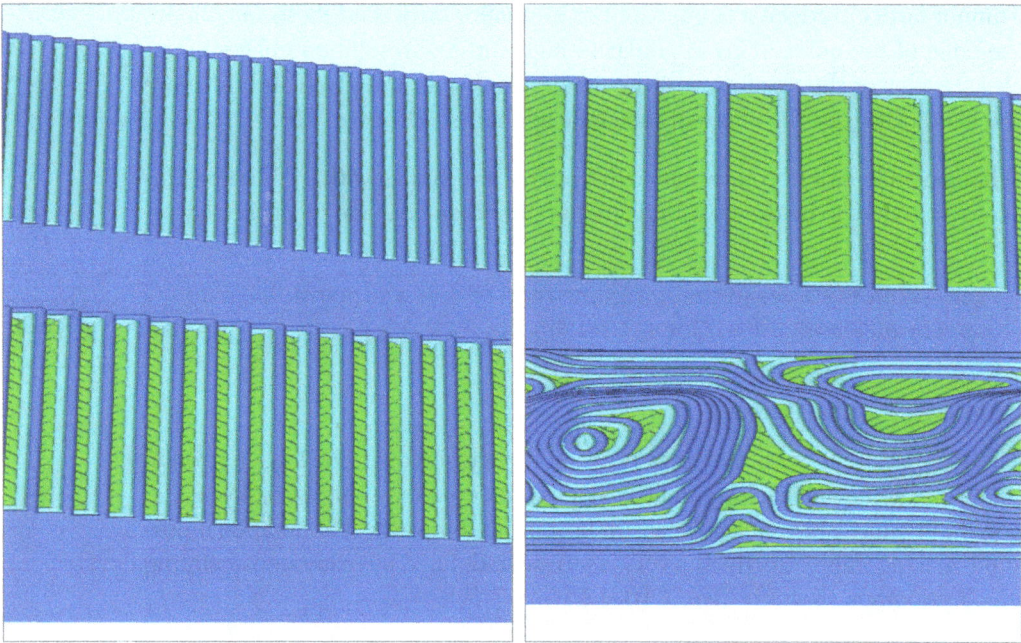

FIGURE 7.11 – Low layer height/high resolution of 0.1 mm with the same objects and orientation. Note that 6 degrees is just starting to be look better; the rest still shows pronounced stepping. The curved space looks much better, showing much more detail but still is quite stepped. Jonathan Torta.

## 0.1 mm

The same objects but printed vertically: even at a high layer height/low resolution, there is a lack of the exaggerated stair-stepping (see **Figure 7.12**).

FIGURE 7.12 – High layer height/low resolution of 0.3 mm printed vertically. Note: these are the same wedges as Figure 7.11 and Figure 7.10. Jonathan Torta.

For the 0.4 mm nozzle, I find that 0.2 mm or 0.15 mm is a sweet spot and good compromise between speed and resolution.

Some slicers allow you to change the layer height dynamically during a print. This is most useful for prints that only have detail in a specific section. I like using it for the curves along the top or bottom of a print and when changing orientation is not possible.

For example, when printing a sphere, I can start out at a higher layer height, decrease it as the curve decreases along the mid-point and then increase it again for the top layers. This reduces the stair stepping that is most noticeable on the very bottom and top of the print and allows the print to speed up in the middle section where vertical resolution is not critical.

## Fill density

### Short Description

Infill refers to the density of the internal space inside the outer shell of an object. This is often measured in percentages (%) instead of millimeters (mm) like the layer height.

### Detailed Explanation

If an object is printed with 100% infill, it is completely solid on the inside. Generally, the higher the percentage of infill, the stronger and heavier the object is, and the more time and filament it takes to print. This can become expensive and time-consuming if you're printing with 100% infill every time – so keep in mind what you'll be using your print for.

- For an object with a large surface area on top, I would generally use a minimum of 18% infill.
- For something to be mechanically strong, I would throw in an extra wall/shell thickness (below) and go up to 40% infill.
- For an item for display, 10–20% infill is recommended.
- For an object that is going to be more functional and sturdier, 75–80% infill can be used.

Different programs can generate different infill patterns inside your object, which can give the top layers of your model more support and add strength, or just look fun (grid, triangle, and honeycomb).

These tend to be strong and give good support to upper layers. Other patterns like wavy lines, random, or sparse are much less useful in almost all cases (see **Figure 7.13**).

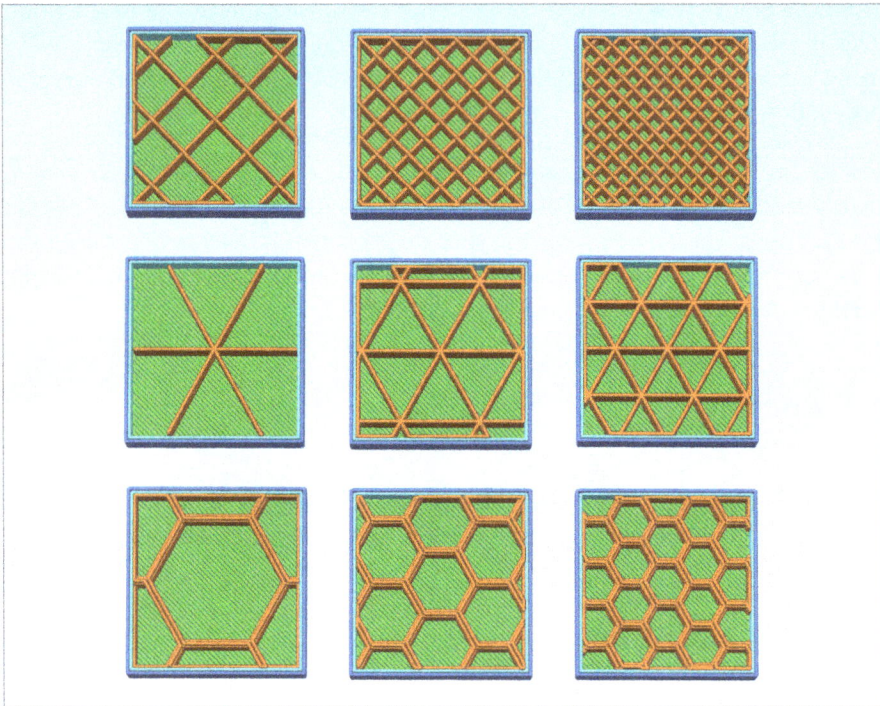

FIGURE 7.13 – Infill patterns. Jonathan Torta.

I have found that Infill is best used for supporting the upper layers. A balance of infill and top layer thickness can be used to keep a quality print and use less material. For strength, while I found that, while infill does add some structure and rigidity, it is not proportional to the infill amount; increasing the wall/shell thickness (below) has a greater overall effect.

## Supports

### Short Description

Supports are printed structures to hold up 3D objects that don't have enough base material (surface area) touching the build plate during printing. This is important to remember since objects are printed in layers. The object and all its parts that do not touch the build plate or extend past a 45-degree overhang angle will have little to no first layer of filament to build on.

## Detailed Explanation

For the print as a whole, supports keep the print on the build plate. If the print falls off the build plate, you have a failure.

For example, during cooling, the parts of an object above the build plate that have no supports (called overhangs) can create a drooping effect, curls, or, at the most extreme, "nests" of plastic without supports, if not adhered well (see **Figure 7.14**).

Most slicers will add supports automatically; some require you to add them manually. Common settings are the angle of the support, to generate supports from the build plate only, or from anywhere ("anywhere" means that supports can actually start on the

FIGURE 7.14 – "Beautiful Failure" close-up; even the failures can be works of art. Photo images courtesy of Jamber, Inc.

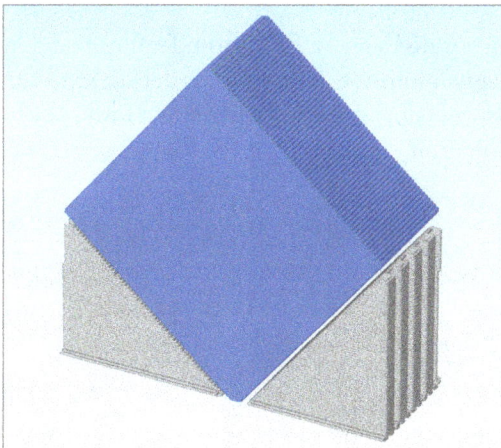

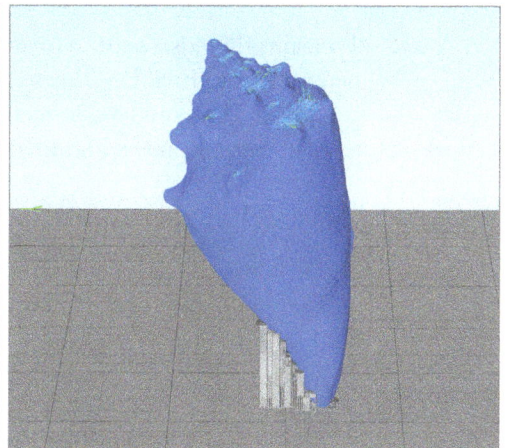

object itself). Still other slicers or advanced settings allow you to edit the support parameters and, even better, to put supports in manually or edit them.

How do you know whether or not your design needs support? The answer is overhangs. Any overhang over 45 degrees needs supports (see **Figure 7.15**).

FIGURE 7.15 – Supports (in gray). Jonathan Torta.

Another way of thinking about it is by using letters of the alphabet.

For instance,

- A "Y" or a "V" is a good shape to print if the base is big enough to hold it in place and the sides are not angled more than 45 degrees.
- A "T" or an "H" would need supports to provide a platform for the horizontal areas. Potential for bridging here.
- An "O" might need internal supports and more around the base (to support the start of the curve and to ensure enough surface area to hold it there).
- An "n" might need support around the top of the curve or it might print ok.

What other letters or symbols can you come up with solutions for?

- "K" does not need any support (if less than 45 degrees).
- "S" would need a lot of support (almost everywhere inside and out) and probably should be printed in another orientation.
- "R" needs support for the horizontal parts.

Also note that different materials and cooling methods can greatly affect the support. PLA can cool quickly with little to no shrinking. This makes it very good for printing at much higher overhang angles and long bridges.

## Platform adhesion – skirt/brim, raft

### Short Description

Typically, there is a setting to enable or disable a raft and brim feature to aid in bed adhesion. Enabling this setting will improve how your model sticks to the print bed. This could aid in models with small footprints (such as a tower) or to combat warping at the bottom of the print.

### Detailed Explanation

#### Skirt/Brim

A skirt is an outline of the bottom layer of your object. Skirts are printed first and are useful to ensure the plastic is primed and flowing correctly before starting your main object.

A brim is similar to a skirt, but is attached to the object and widens the print's bottom footprint. Like the base of a floor lamp, it creates more surface area to stick to the print bed. Since it is normally only a few layers (often just one), it is relatively easy to remove. These wide brims can keep the corners of your model down without leaving marks across the entire bottom of the object.

## Raft

A raft is a horizontal grid on the print bed that acts as a platform to help adhere the print to the bed. It aids greatly with print beds that are not perfectly flat, as the print will start on the raft rather than the bed, and the raft's first layer is printed quite thick, allowing quite a lot of tolerance. This can also be useful when printing objects with many small parts. Remember that a raft sticks to the object and will need to be removed. This may roughen the surfaces touching it.

# Number of shells/perimeters

## Short Description

Shells/perimeters refer to the number of times the outer walls are traced by the 3D printer before starting to print the infill sections of your design. This applies to the sidewalls only. This can also be called wall thickness. Wall thickness may sound easier to use and visualize ("I need a 1 mm wall") but remember, the sides of the print are dictated by your nozzle size, not the layer height. So, with a 0.4 mm nozzle, a setting of a 1 mm wall is not quite two and less than three shells, as each pass with the nozzle lays down a 0.4–0.48 mm line of filament. So, unless the wall thickness is a multiple of the nozzle size or, more importantly, the extruded line, it will not print to the thickness you set it to. That is why shells are often used, as they are by definition a multiple of the nozzle size.

## Detailed Explanation

This number defines the thickness of the sidewalls (the X and Y dimensions) and not the top and bottom. It is commonly between 0.8 mm and 1.2 mm for a 0.4 mm nozzle, or two to three shells. Different slicers use can different units – millimeters, shells, or both. It can be set thinner for decorative prints and thicker for prints requiring strength.

This number is one of the biggest factors in the strength of your print. Increasing it will create thicker walls and greatly improve the strength of the print.

To handle the top and bottom of the print not controlled by shell thickness, most slicers separate the top and bottom layer settings, since they are handled by the z- axis, but will adjust accordingly with basic settings.

If you set the advanced settings separately, you need to take into account the layer height. If you were to have two shells with a 0.4 mm nozzle (~0.8–0.96 mm wall) then you probably want to have top and bottom layers set to a similar thickness. For example, 0.1 mm layer height needs 8–10 layers, 0.2 mm 4–5 layers, 0.3 mm ~3 layers. This will make the top and bottom of the print similar in thickness to the sides. If your slicer does not automatically do this for you, you should always have 3+ layers minimum for a good top and bottom.

**MAKER'S NOTE**

Some slicing applications offer additional basic controls such as retraction, print speed, initial layer thickness, bottom/top thickness, and spiralizer.

## › 7.5 – What are some common advanced slicer settings?

In addition to the basic settings, there are a number of advanced settings. We will examine these settings in more depth in Chapter 9 when we discuss setting up slicing parameters. However, here is a quick list for reference.

Commonly used/modified advanced settings:

- Extended extruder settings
- Shell/wall thickness
- Extended infill settings
- Extended support settings
- Fine control of temperature
- Fine control of cooling fan
- Fill pattern and density
- Speed settings
- First layer settings
- Bridging settings
- Temperature Settings
- Ooze control
- Prime pillar and ooze shield

Special modes or experimental settings:

- Vase mode
- A draft shield
- Multiple setting profiles or dynamic settings
- "Fuzzy" exterior
- Ability to print multiple objects on the print bed one at a time
- Manual or alternate support style and placement

## › 7.6 – Are there additional software applications used in 3D printing?

In addition to 3D model creation and slicing software you might use additional applications for transferring slicer file information to your printer and monitoring your 3D printer during print. We will examine transferring and monitoring in Chapter 9; however, below is a list of applications that could be used for both transferring and monitoring.

## Transferring data

After you export the sliced model into a file format your printer can use, you need to transfer the information to your printer.

There are a few ways to send your files to the printer:

- **SD card** – Copy file and insert into printer manually
- **Dedicated Controller (Octo-print/Rocketprint/computer)** – USB streaming to printer
- **Wi-Fi** – Streaming to printer or copying file to printer or controller.

## Monitoring

After your printer has started a print, you might want to oversee the printer during operation (see **Figure 7.16**).

FIGURE 7.16 – AstroPrint interface on a tablet while remotely managing a 3D Printer. JayLoerns

## Common Monitoring Methods

- Octoprint/AstroPrint – Controls and monitors every aspect of a 3D printer
  › Browser controlled
  › Embedded webcam feed
  › Gives constant feedback regarding the current progress of your print job
  › Monitors the temperatures of your hot ends and print bed
  › Moves the print head along all axes, extrude, retract
  › Start, stop, or just pause your current print job
  › Push notifications (texts, emails, etc.)
  › Mobile device apps
  › Collect statistics
  › And much more
- Baby monitors – Simple remote viewing of printer

- Remote power plugs (to cut power) – Various smart plugs allow control by mobile device apps, or even voice control with Google Assistant or Alexa
- Various other mobile device apps (Printoid app, 3D Fox app, easy print app, and other proprietary apps)

## SUMMARY

There is a wide range of software applications for slicing using a 3D printer, from beginner to industrial and from no cost to high cost for commercial use. When choosing a slicing program, there are five main features to make sure the program has besides compatibility with your 3D printer. Overall knowledge of the settings in a slicing program is needed to produce the desired 3D object design.

## APPLYING WHAT YOU'VE LEARNED

1. Continue making your own 3D dictionary by adding the definition (in your own words) of five words related to 3D printing in this chapter.
2. Define the following terms in your own words as related to 3D printing: file formats, features, complexity, operating system, and overall cost.
3. Go on the internet and find additional information about software for 3 D printers creating and editing objects.
4. If you ae using a 3D printer or scanner, what level are you and which software programs can you use for your printer, including information such as type, operating system, cost, and readable formats?
5. Add five more dedicated object creations to the list in this book and explain what they do.
6. In your own words, explain the five common basic slicer settings.
7. Explain the five common advanced slicer settings.
8. Name the five extruder settings and discuss them.
9. Explain why are infill settings are important and give an example of how an incorrect setting would affect the item.

## REFERENCES

[1] – https://www.3dnatives.com/en/3d-software-beginners100420174/

By microgen

# Getting Started

## OVERVIEW AND LEARNING OBJECTIVES

**In this chapter:**

- 8.1 – How do I obtain a 3D Model?
- 8.2 – What are some of the differences between file formats?
- 8.3 – How do I condition or repair a 3D Model for slicing?
- 8.4 – When do I slice my 3D model?
- 8.5 – Should I use a 3D printing service?
- 8.6 – What do I look for in a printer?

Before you can start a 3D print, you need to make some preparations. This chapter focused on the beginning process that we briefly discussed in Chapter 1, which includes: how to obtain a 3D model, checking or fixing the model to see if it is ready for printing, and what file formats best fit your project.

After your 3D model is ready, you will need to choose the method you will use for your print, like using a 3D printing service or store, buying or building your own 3D printer, or a combination of both methods.

## › 8.1 – How do I obtain a 3D Model?

Obtaining a 3D model is your first step. There are a few ways to do this, including: building a 3D model from scratch, scanning an existing physical object, or using an existing model you find online.

These methods can be used individually or in combination. For example, you can download a 3D model from an online site, then bring it into a 3D modeling application to make modifications before you print.

### 3D modeling

You can make your own 3D model using one of the many CGI and CAD applications on the market. These include animation, art, gaming, or CAD applications. Each application has a different set of features. Some of these are geared toward beginners and others are more complex. See Chapter 7 for more information on the types of applications.

Making your own model can be quite fun and some applications have made it much easier to create. Try out a few applications and find one that works for your skill level and budget.

### 3D scanning

Even though you can use 2D cameras to reconstruct and render an image into a 3D model, a dedicated 3D scanner will scan an object directly into a 3D model. (See **Figure 8.1**.)

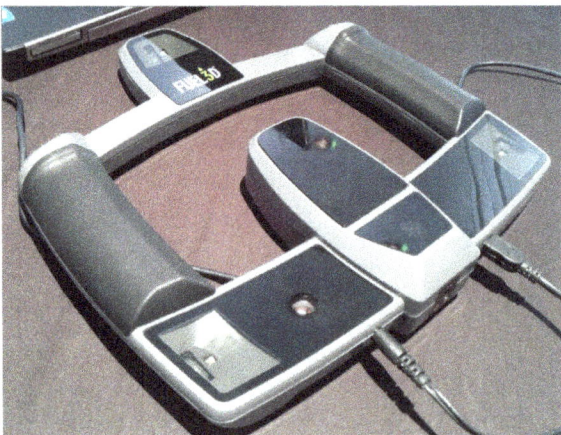

FIGURE 8.1 – Fuel3D portable 3D scanner. Creative Tools.

3D scanners can be very useful tools in your 3D printing workflow. If you have a pre-existing object, you can scan it to create the 3D model. You can then prepare the 3D model for printing or make modifications to it.

3D scanners come in all shapes and sizes and can be tailored to how they are being used. For example, some scanners can scan whole rooms while other are made into work desks.

## Parametric methods

Parametric modeling is a subset of 3D modeling that uses mathematical equations to create 3D models. Although not as popular as other methods, the parametric method is sometimes used by online sites. An example is seen in **Figure 8.2**, which shows the snippet of code that was used to create the adjacent image.

```
set term povray
set output "~/prj/wikipedia/reebfoliation/torus.pov"
set view 40,280
set parametric
set iso 15
unset key
unset border
r=3
fx1(u,v)=(pi/2*sin(v)+r)*cos(u/(pi/2-0.3)*2*pi)
fy1(u,v)=(pi/2*sin(v)+r)*sin(u/(pi/2-0.3)*2*pi)
fz1(u,v)=pi/2*cos(v)
fx2(u,v,n)=(abs(u)*sin(v)+r)*cos(1/cos(u)+2*n*pi/6)
fy2(u,v,n)=(abs(u)*sin(v)+r)*sin(1/cos(u)+2*n*pi/6)
fz2(u,v,n)=abs(u)*cos(v)
splot [0:pi/2-0.3] [2*pi/4:6*pi/4] \
fx1(u,v), fy1(u,v), fz1(u,v) w l lt 6, \
fx1(u,v), fy1(u,v), fz1(u,v) w d lt 6,\
fx2(u,v,0), fy2(u,v,0), fz2(u,v,0) w l lt 2, \
fx2(u,v,0), fy2(u,v,0), fz2(u,v,0) w d lt 2,\
fx2(u,v,1), fy2(u,v,1), fz2(u,v,3) w l lt 3, \
fx2(u,v,1), fy2(u,v,1), fz2(u,v,3) w d lt 3,\
fx2(u,v,2), fy2(u,v,2), fz2(u,v,3) w l lt 2, \
fx2(u,v,2), fy2(u,v,2), fz2(u,v,3) w d lt 2,\
fx2(u,v,3), fy2(u,v,3), fz2(u,v,3) w l lt 3, \
fx2(u,v,3), fy2(u,v,3), fz2(u,v,3) w d lt 3, \
fx2(u,v,4), fy2(u,v,4), fz2(u,v,12) w l lt 2, \
fx2(u,v,4), fy2(u,v,4), fz2(u,v,12) w d lt 2, \
fx2(u,v,5), fy2(u,v,5), fz2(u,v,15) w l lt 3, \
fx2(u,v,5), fy2(u,v,5), fz2(u,v,3) w d lt 3,\
fx2(u,v,6), fy2(u,v,6), fz2(u,v,15) w l lt 2, \
fx2(u,v,6), fy2(u,v,6), fz2(u,v,3) w d lt 2,\
fx2(u,v,7), fy2(u,v,7), fz2(u,v,15) w l lt 3, \
fx2(u,v,7), fy2(u,v,7), fz2(u,v,3) w d lt 3
```

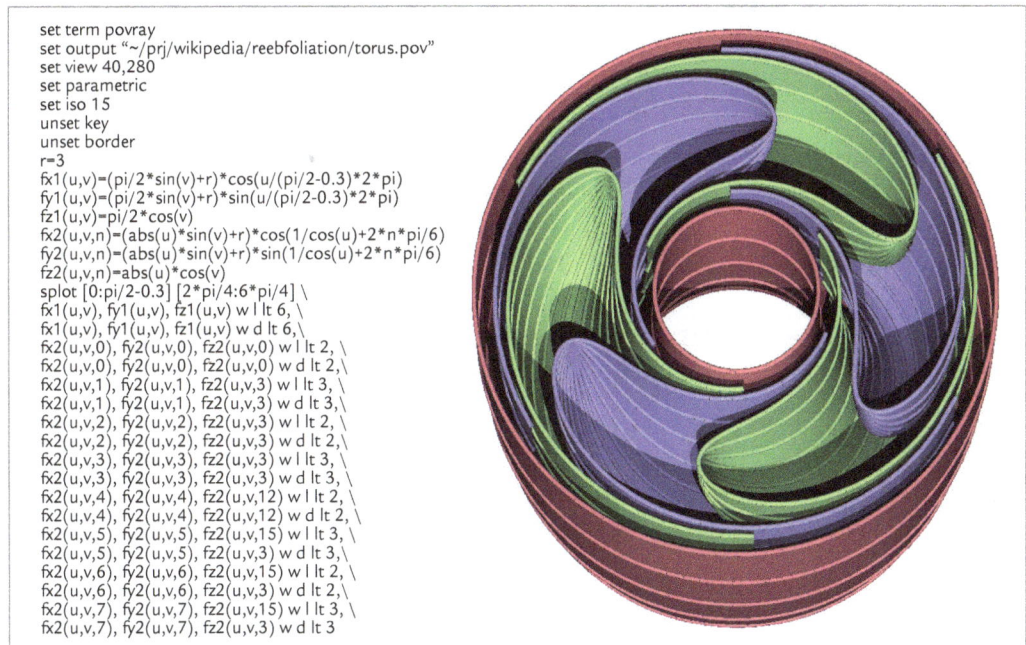

FIGURE 8.2 – Reeb foliation, generated with Gnuplot and POV-Ray. Ilya Voyager.

## Online downloads

One of the easiest ways to obtain a 3D model is to download one from an online repository. Premade 3D models can be downloaded from the internet for commercial and private use. Some sites have many categories and types of objects, while others focus on niche markets (see **Figure 8.3**).

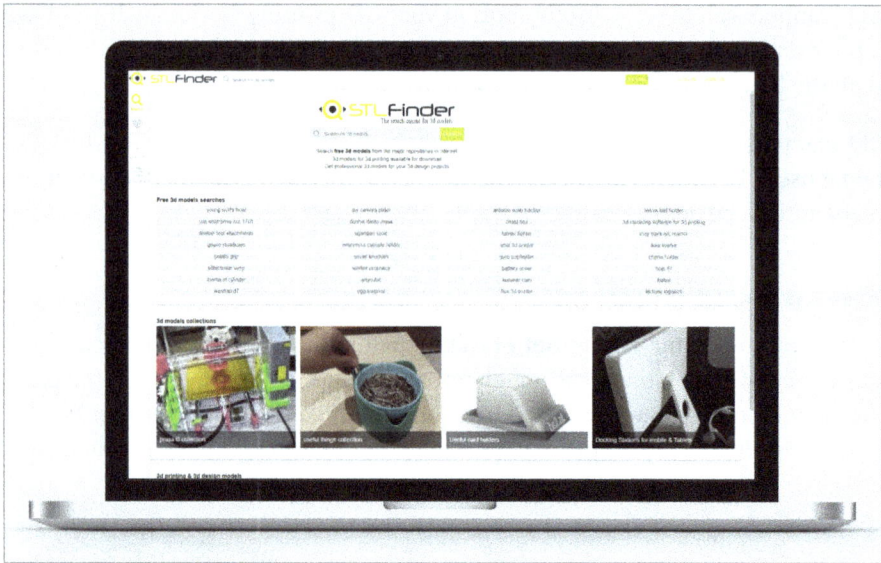

FIGURE 8.3 – An example of a 3D web site for downloading 3D models.

There are also web sites that allow makers to upload and download freely with collaboration and version enhancement. We will talk more about the community and collaboration of online sites in Chapter 15.

## › 8.2 – What are some of the differences between file formats?

As stated in Chapter 1, stereolithography (STL) is the most common format. You'll likely encounter this file type when searching online for models. However, there are a few other file formats that are being use and some that are growing in popularity.

Some advantages and disadvantages of the selected file formats:

**3D Manufacturing Format (3MF)** – A relatively new format that can store model information like shape, color, and material. This is handy for multicolor printers. It has everything needed for 3D printing but does not have wide adoption as of yet.

**Additive Manufacturing File (AMF)** – A relatively new format that can store model information like shape, color, and material. This is handy for multi-color printers. These can be in XML text form or in a compressed form that is about half the size of the Binary STL size and has everything needed for 3D printing but a wide adoption rate.

**Object file format (OBJ)** – A common open source format that can contain additional information like UV mapping and texture coordinates. This format is made for 3D printing but is common and freely useable. It is most often an export format between programs.

**Stereolithography (STL)** – The most common format, which comes in two versions (binary and ASCII). It is a very simplistic format containing only the mesh and normal file information. The coordinates have no units, but typically use millimeters or inches, and you have to specify them (for example, if the object appears too small, change the units to millimeters, unfortunately as there is no way for the program to know which to select so this is setting is manual). This format tends to have the largest file sizes.

## › 8.3 – How do I condition or repair a 3D Model for slicing?

Not all 3D models come ready to print. Sometimes the 3D model will need fixing before it can be used to slice for 3D printing.

The 3D models we will be using are created by defining the surface of the object. For 3D printing, it is required that the object be *manifold* (or watertight). Essentially, the 3D model is not solid; it only has a skin defining the exterior and can be seen as hollow. If filled, it should be able to hold water.

A problem occurs when the object is not watertight and does not have all of the edges continuously connected. Any problems with the geometry can confuse the slicer and give poor results or make it completely impossible to print.

> Not all 3D applications care much about the objects being perfectly manifold, as most just care if they look good on screen, where small unseen flaws are no issue. With 3D printing these issues become very important.

**MAKER'S NOTE**

Some key object quality factors that need to be checked:

- **Manifold:** Manifold models have a well-defined interior and exterior (think watertight interior between inner and outer surfaces). This is defined by the object's geometry, and any damage to the object's mesh can cause the object to be no longer manifold.

- **Model geometry issues:** Overlapping faces, duplicate triangles, and intersecting geometries break the manifold and are generally considered broken.
- **Open Boundary edges:** Holes in the surface mesh and open edges also break manifold.
- **Flipped faces:** Every surface/face has one side. This is defined by the normal of that surface (a perpendicular line from that face). If the normal is pointing to the interior of the object, the face is considered "flipped." Note that this also helps define whether the object is manifold.

A common process is to check an object for quality after downloading or first use of an object. Some slicers have automatic checks – Meshmixer, 3D Builder, MeshLab, Blender, Open3mod, MeshFix, and Autodesk Netfabb – some slicers fix problems, and still others have no checking at all.

There are a few online services that perform the checks, including: makeprintable.com, netfabb.com, tools3d.azurewebsites.net.

> **! TIP**
>
> It is always a good policy to check, as there are many malformed STL files that are shared. You can still use most of these files, but they might need work. Some automatic fixes might not work completely. You may have to fix them manually. It is always a best practice to determine the quality of the 3D object's mesh before printing.

## › 8.4 – When do I slice my 3D model?

After your 3D model object is in good condition to print, the next step in the process is to slice your object into printable layers and generate a tool path and instructions for the printer. See Chapter 7 for more information on free and purchasable slicing software.

The slicing program, in accordance to your settings, slices the 3D object into layers, adds supports where needed, and makes infill patterns for the interior of the object. The 3D printer then uses this information to print the object layer-by-layer. **Figure 8.4** is an example of a 3D object within a slicing program.

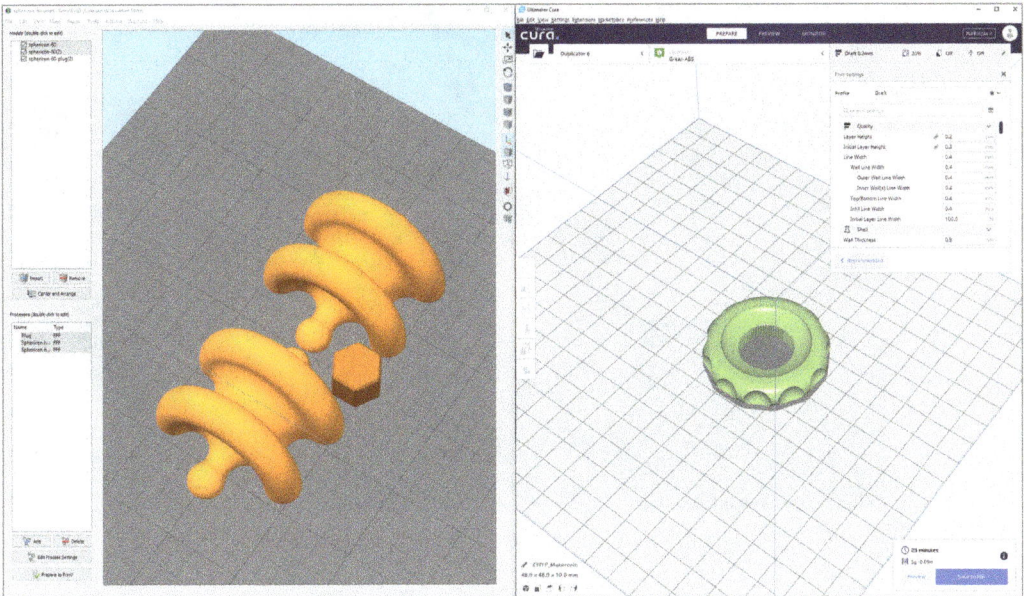

FIGURE 8.4 – Samples of 3D models in slicing programs. Jonathan Torta

More specifically, the slicing software converts the object and printer settings into thin layers as a toolpath for your extruder. This tool path and other printer information is compiled into G-code that the 3D printer follows. This G-code carries all the instructions that the printer needs to print the object, like temperatures (hot end and bed), all movements, and fan speeds, etc.

The slicer program may fail to slice an object correctly. Any errors can affect the quality of the slice. We will be taking a closer look at the types of errors and what causes them in Chapter 10.

## › 8.5 – Should I use a 3D printing service?

If you do not have the time, space, or opportunity to buy or build your own 3D printer, there are physical stores, online services, and individual DIY makers you can utilize to print your project. (See **Figure 8.5**.)

One of the major advantages to using one of these 3D services is that many offer a wide range of print methods and printable materials. Some of these services offer 3D model creation and post process to finish your object.

For example, the online service Shapeways.com can print objects in materials that are not available to most DYI enthusiasts.

FIGURE 8.5 – 3D Printing Dublin, Ireland's first 3D Printing store. CLS500.

Some of these materials include:

- Steel
- Platinum
- Gold
- Silver
- Bronze
- Brass
- Aluminum
- Sandstone
- Professional or fine detail plastic

**Web link** Shapeways.com is an example of a full range online 3D printing service: www.shapeways.com/create

We hear from Tony Hu about his experiences in using a variety of 3D printing services.

**TONY HU** **QUOTE**

*Academic Director, Integrated Design & Management Program, MIT; Chief Yak, Brainy Yak Labs*

www.brainyyaklabs.com

Service bureaus are handy because they allow you to choose from a wide range of technologies and materials without investing in equipment. You have the freedom to select between a quick FDM, a transparent SLA, and a refined Polyjet, depending on your needs. You can also shop around for the best price or turnaround time. I've paid a bit more to a shop in Canada in order to use a specific machine and material. I've also used a local source for a quick proto – a young guy showed up with my part at the door to his apartment while wearing pajamas.

Another advantage in using online services or physical stores is that you do not need to buy the equipment. Individuals or businesses can take advantage of manufacturing equipment that best fits their project without having to buy or learn the equipment. We talk with Harnek Gulati about his online factory.

## HARNEK GULATI

*Position and place of work: CEO of MakerFleet*

makerfleet.com

**ST: What is MakerFleet?**

HG: MakerFleet is an online factory that allows users from around the world to directly access manufacturing equipment in our factory.

Makerfleet is not a manufacturing company, but an online factory, the first of its kind. You never need to own any manufacturing equipment – whether it's a 3D printer, a laser cutter, or a CNC – to start using one. You can learn how to use the machine without needing to own one. And the more you learn, the more you can make. (See **Figure 8.6**.)

**ST: How does the process work?**

**HG:** People upload their model files and place them onto a print bed, just like the normal 3D printing process. After slicing, they are directly connected to a 3D printer and can watch the stream of their 3D print. You can cancel the print if something goes wrong. We record the time-lapse for later sharing and inspection.

**ST: Does a consumer need to understand 3D printing to use MakerFleet?**

**HG:** We have a beginner's guide that gets people started, but our software also makes the whole process really easy (see Figure 8.7). You can get from "I know nothing" to "I'm starting a 3D print" in less than five minutes. We also have plenty of guides online that allow people to improve in 3D printing, all without needing to own their own 3D printer.

FIGURE 8.6 – A bay of 3D printers. Harnek Gulati.

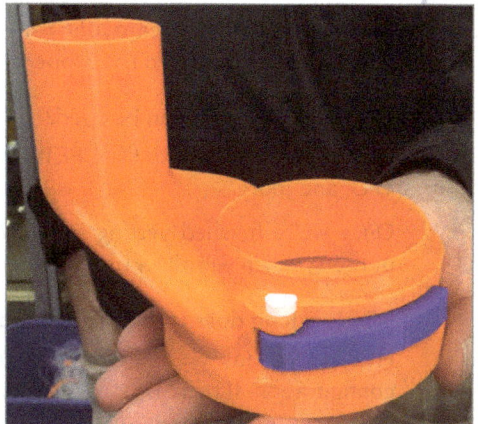

FIGURE 8.7 – A complex printed part. Harnek Gulati.

**ST: What does the consumer need to start with?**

HG: You just need an internet connection. We provide you with an example STL file.

**ST: What objects can be printed?**

**HG:** All things an FDM printer can handle. To be a little more specific, FDM printers can't handle thin wall thickness, highly detailed objects, and objects that require quality on some overhang angles. Our software detects most of these corner cases and will let you know how to change your prints. (See Figure 8.8.)

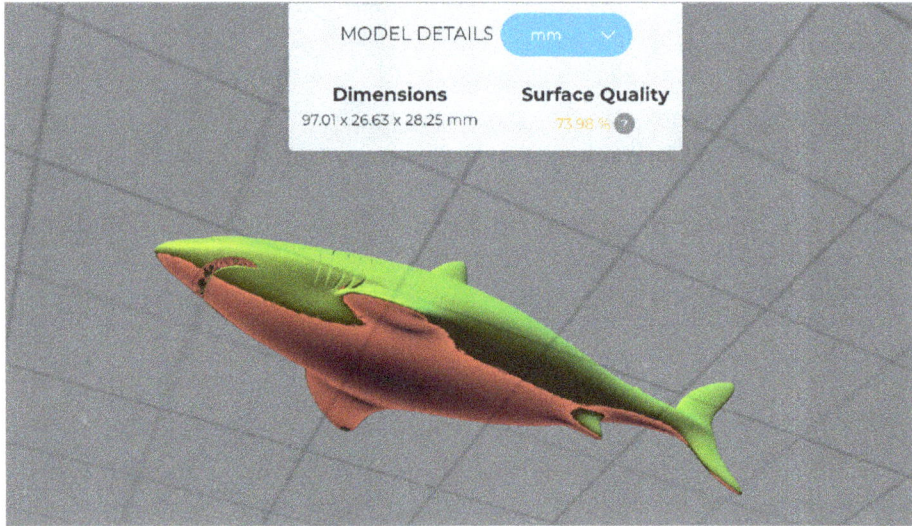

FIGURE 8.8 – A 3D model being checked for errors. Harnek Gulati.

**ST: How big can the prints be and what quantities can be printed?**

HG: Currently, they're limited to be within a size of 210 mm x 250 mm x 250 mm.

Anything you can dream of. You can order 500 objects that our factory will automatically load onto the proper 3D printers.

**ST: What materials can be used?**

HG: We're starting with PLA, PETG, and soon will be moving into carbon fiber nylon, TPU, and more.

Once we're finished getting our FDM printers off the ground, we'll be using SLA printers, which will allow us to get better prints.

**ST: How long does is the turnaround time?**

HG: It depends on if there is an available 3D printer in the factory with your configuration. If there isn't, we'll tell you how long before the next one is available.

**ST: Are there any project or stories you would like to share in about your 3D printing online factory?**

HG: People have been using our online factory for all kinds of projects (see Figure 8.9):

> **Popsicle mold makers:** A company came to us asking to use the 3D printers. We helped them design their mold makers.

> **Medical Devices:** A medical device company extensively used our printers to get started on building things.

> **Bones:** The Harvard Edx Portal wanted to print out bones to help students learn about the physical approximations.

> **Cosplay:** We have people using us to create their cosplay.

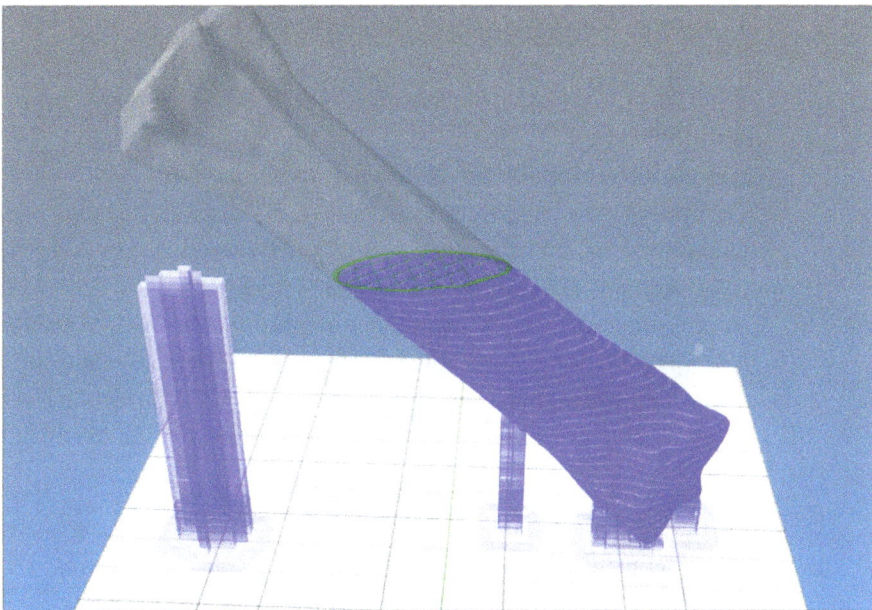

FIGURE 8.9 – A bone 3D model in slices with supports. Harnek Gulati.

Another advantage can be quick turnaround times and convenience. Services often have bays of 3D printers ready for use. This can save time if you are printing larger quantities, as more than one printer can be working on the same project.

3D printing services range from full service to specialized printing. Some DYI enthusiasts have even started their own side businesses, offering to print objects for consumers. Even if you own a 3D printer and print objects regularly, there still are times where a 3D print service may fit the needs of your project more efficiently.

## › 8.6– What do I look for in a printer?

When looking for a printer to buy or build, you first need to decide on a few factors to help narrow down the selection. This book covers the common factors to consider before selecting a printer.

Some common factors include:

- **Size** – Printers come in all sizes, from 12-inch cubes to monsters that are many feet in size; most printers are small, from 100–250 mm square, but this can vary quite a bit
- **Price** – 3D printers range from $200–$5000 and average around $500–$1000 for a "full featured" one
- **Features** – There is a long list, but features that are big now include auto-tramming and multi-color
- **Print area** – Maximum and average size you can expect to print with the printer
- **Capability** – Material compatibilities, minimum layer height, speed
- **Complexity** – Some printers are made to be very simple to use; others are known for their advanced features and larger learning curves
- **Availability** – 3D printers and filaments are a worldwide market and some models are harder to purchase in other regions; lack of availability and language barriers can affect the quality of instructions and support
- **Cost of supplies and tools** – you need to purchase more than just the printer: filament, tools, and even workspace items like tables and lighting

## Buying or building a 3D printer

Before you buy a desktop 3D printer, there are a few questions you should resolve.

- What are you going to use the printer for?
- What style/type of printer are you looking for?
- What is your budget?
- Do you want to modify your printer?
- Do you need a fully functional, out-of-the-box system with service support?
- Do you need high reliability?
- What space do you have available?

- What size printer area are you looking for?
- What materials do you need to print in?
- What features are must-haves?
- How important is having a printer with multiple heads or multiple color printing?
- How often do you intend to print?
- Is there community support and existing modifications (mods) available?
- Are you a DIY-modder type that likes upgrading things, or do you see the printer as an appliance that should just work?

Once you answer these questions, you can narrow down the list of printers to choose from.

## Two printer categories for the home user

### Ready-to-use or Consumer Level Printers

These tend to be slicker, ready-to-go printers bought in a common storefront with extensive features, instructions, and software, ideally made to hold the user's hand throughout the setup and printing process. They can have proprietary hardware and software and a limited ability to be modified (if at all) to extremely basic designs. They are the upper end of this category, with all of the convenience, automation, and support, and they tend to be relatively expensive (the best are $1000 and up). However, if you want a no-hassle experience, this may be the best option to make 3D printing as easy as possible. At the lower end of the category it is truly "buyer beware." Many of the cheapest printers need extensive calibration just to print, may print well for a short time and then need extensive maintenance, have little or no instructions or support, and are generally a nightmare for the average consumer who just wants to print with simplicity and ease of use without breaking the bank. These printers also tend to be the cheapest knockoffs, using substandard parts and cutting corners. And while you may find a gem, it is often the luck of the draw. So, if you do not feel you can troubleshoot and replace important parts (including most of the hardware and electronics) yourself, you need to stay away from that end of the spectrum.

We hear from Zac Sund, who bought a ready-to-use 3D printer for problem solving and hobby projects.

## DIY printers

These are the mainstay of 3D printing. Hundreds of small companies are building new versions of 3D printers that compete for the ultimate low priced, modestly featured, very modifiable printer for the DIY enthusiast. A few high-end models are opensource and full featured. And some are in kit form and use common parts, so even building from scratch is quite possible with the wide range of hobbyist hardware, open source software, and firmware resources available.

Alternatively, the DIY group may very well use cheap "consumer" printers (the ones average consumers should stay away from) as base units to modify and upgrade into something more useful and reliable. They love the idea of replacing parts, tuning, and generally overhauling their printer before using it regularly.

A large portion of these printers are in the few-hundred-dollar ranges, and most are below $1000. This is a good path if you want to not only do 3D printing but also want to build or modify a printer and learn its inner workings. Note that what you may save in initial cost is often made up in effort, learning, and modifying and buying more components.

We will go more into depth on modifying your printer in Chapter 14.

**MAKER'S NOTE**

In Chapter 14, we list online references and keyword searches for further research on the right printer to buy or build that fits your needs, along with modification possibilities.

There is always a learning curve when buying your first 3D printer. We next talk with Dale Hawley for his firsthand account as a beginner after buying his own printer.

## Firsthand Account: Beginner Buying

### DIY Hobbyist: Dale Hawley

I came into 3D printing fairly raw. I knew a few things like the fact that good 3D prints take a lot of time and that there were resolution limits, especially with PLA filament, and that there were tons of resources out there, which actually became part of the problem but I'll address that later.

Once I did finally select a starter-printer, I quite literally spent the next full week discovering what I didn't know, which was both interesting and unbelievably frustrating because information overload quickly set in and it started to feel like I was never going to actually get to printing anything beyond the "Benchy" test prints. I'm pretty sure that is the point at which a lot of people just throw their hands up in frustration and say "this isn't worth it."

Setup itself was an adventure. I knew that setting nozzle height and leveling the bed were critical, however there were other things I discovered that I wish I had been prepared for ahead of time. Learning these facts "on the fly" contributed a lot of stress to the learning curve.

Such things as:

1.  When you set up a new printer, you really must take a true straight-edge and make sure the bed is flat! Even a warp or bend that is too small to see can create huge headaches for the new 3D printer user because you are working in microns, so a small warp is a BIG deal. For a heated bed, check level once the bed is warmed up to temperature.

2.  Because of how slowly things print, "dialing in" a printer can take a LOT of time and consume a fair amount of filament before you get things just right. Initially I didn't set aside enough "quality" time for this and had to keep coming back to it, increasing frustration and creating a pile of failed garbage.

3.  Ambient temperature MATTERS! There is a reason a lot of people create enclosures to regulate temperature and drafts. I wish I had known ahead of time to allocate space and materials for an enclosure, not just for the printer itself. In short, even a small 3D printer can wind up taking quite a bit more shelf/desk space than anticipated (see **Figure 8.10**).

4.  Buy a few spare parts along with the printer: a spare nozzle or two, some PTFE tubing, a few replacement tube fittings. You may not need them, but it is super frustrating to be on the learning curve and get tripped up by a clogged nozzle that you don't know how to clean yet.

5.  Software (and the configuration) matters at least as much as the printer itself.

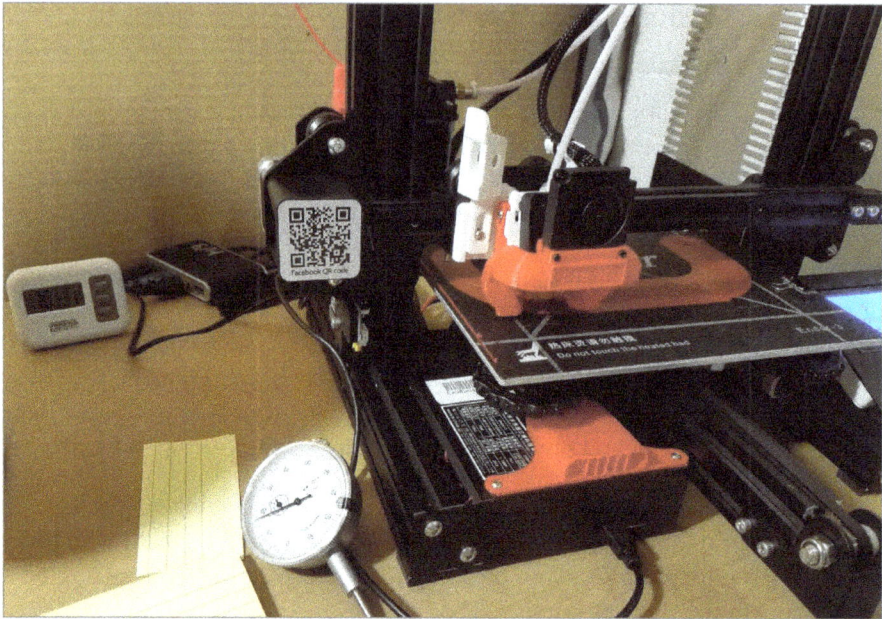

Figure 8.10 – 3D printer setup with temperature gauges. Dale Hawley.

In photography, there are some very well-developed charts that give you a starting point for good photography. It was a bit of a struggle finding consistent information as to what speed/temperature/etc. was a good starting baseline.

One of the things that really surprised me was this focus on "single object printing" that seems to run rampant in the 3D community. While I agree it's neat to create "captured ball" or enclosed gear prints, I've been building models for far too long to worry about needing to assemble multiple parts. It would be really nice if there was software out there that could analyze a 3D model file and give a "score" of difficulty to it so you'd know that "hey, I'm a beginner, I might not want the frustration of trying this level of print until I learn more."

Since I do fine-scale modeling, I would eventually like a printer capable of very fine detail on a very small scale (battleship guns, anchors, all sorts of that stuff). Saw a multi-part 3D build at a local plastic model competition, realized the potential right there. Even with the cost of the printers, in the long run it will be far cheaper than purchasing all of the fine-scale model detail kits, as I've done in the past. The 3D printer also gives me unlimited ability to print in different scales, whereas previously I was constrained by what the companies like MPC or Revell produced.

## SUMMARY

There are four main ways to obtain a 3D model: modeling, scanning, parametric methods, and online downloads. Not all 3D models are appropriate to print. Key object quality factors need to be checked before printing. The best file format to select depends upon what you are printing, because each has its advantages and disadvantages. If you are going to buy or build a 3D printer, there are many questions you must ask yourself before you decide.

## APPLYING WHAT YOU'VE LEARNED

1. Continue making your own 3D dictionary by adding the definition (in your own words) of five words related to 3D printing in this chapter.

2. Describe all the ways you can obtain a 3D model. Which would you like to use and why?

3. Why should the manifold, model geometry issues, boundary edges, and flipped faces be checked?

4. What are the advantages and disadvantages of the following formats: 3EMF, AMF, OBJ, and STL?

5. Find 3D printing services in your area and write/share about them including their services and fees.

6. Summarize the interview by Harnek Gulati in a couple paragraphs.

7. Pretend you are going to buy a 3D printer and answer the ten questions listed in this chapter to help you pick out the right one.

8. If you were to buy a 3D printer, would you buy a read-to-use printer or a DIY printer? Why?

# CHAPTER 9

# Setting Up to Print

## OVERVIEW AND LEARNING OBJECTIVES

**In this chapter:**

- 9.1 – How do I slice my 3D model?
- 9.2 – Do I have to worry about safety and ventilation?
- 9.3 – Why do I need to run hardware tests?
- 9.4 – How can I transfer my file to my printer?
- 9.5 – What happens during a print? Is there anything I need to watch out for?
- 9.6 – Are there ways to remotely monitor my printer?

Now that you have a printer and a model, it is time to get ready to print. Your basic tasks should include:

1. In the slicing application – loading the model, positioning the object, setting up the print parameters and slicing the model.
2. If you have a printer workspace, setting up ventilation and other safety measures.
3. If you own a printer, running additional hardware tests as needed.
4. Transferring the files to your printer.
5. Monitoring the print for progress and any problems.

## › 9.1 – How do I slice my 3D model?

### Slicer applications

After you have a model (downloaded or self-created) and it has been verified as acceptable, load it into your slicer application.

Most slicer applications display the "print volume" (3D space) in which you print. Others may have a profile for your printer that displays a 3D representation of the print platform. This allows you to position the objects as needed inside that volume. They may also allow you to adjust the orientation and size as needed (see **Figure 9.1**).

### Loading Your Object or Project Files

Typically, you load a model like any other file in any other application (see acceptable file formats in Chapter 7 and Chapter 8). Some applications allow you to drag and drop the file into the application. Also note that some applications include project files that save the objects plus the settings, especially advanced settings, variable settings, custom support, and filament used, so you can come back to it later. Several good examples are Simpify3D's "factory" project files and the current Ultimaker Cura using 3mf format to hold extra printer information.

### Positioning the Objects

You should place the object onto the print platform, as printing must start on the print bed. Note that if you use a raft or support, it will be automatically inserted beneath the object. You should find a flat surface on the object and make that your base. A flat area is perfect for a solid surface to stick to the print bed.

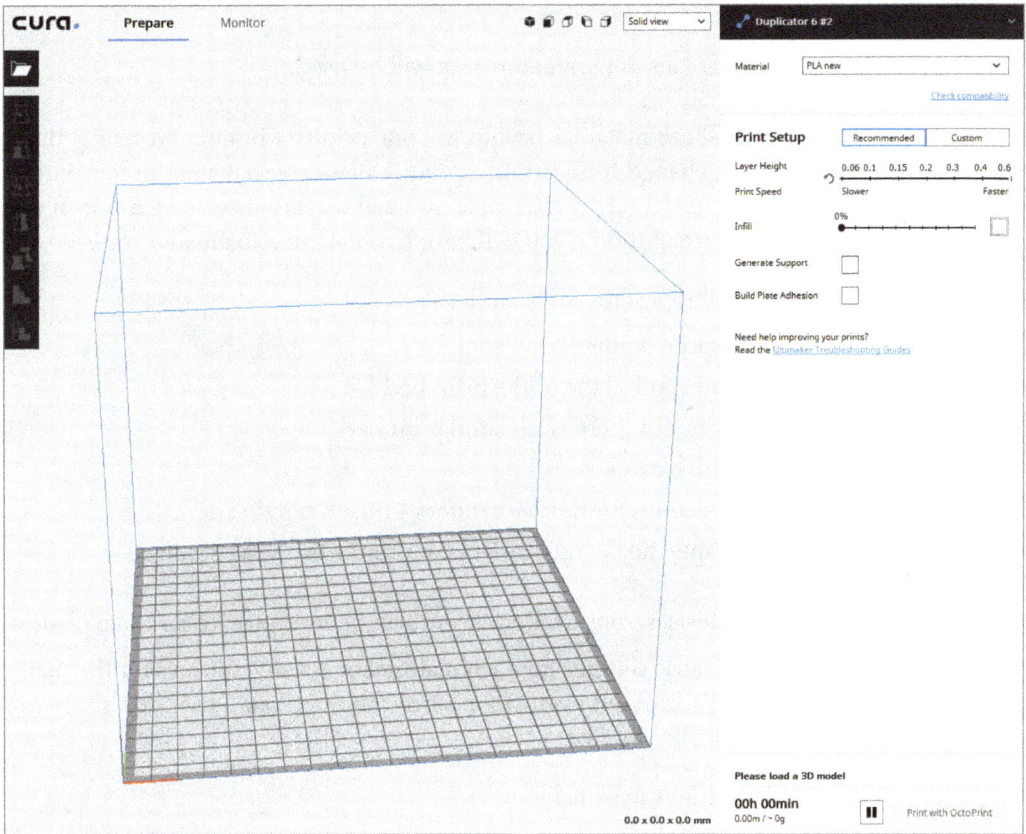

FIGURE 9.1 – Ultimaker Cura Slicer

Common slicers will "drop" your object onto the virtual print bed and either keep the lowest part of the bed or allow you to re-drop the object as needed. This is to ensure that, no matter the orientation of your object, part of it will always be on the print bed.

Some slicers have additional controls and features that can automatically, arrange multiple objects, or to duplicate objects.

> Orienting the object correctly is **very** important, as you need enough surface area to adhere the print to the bed and stick there well. If the object moves or comes loose, your print is ruined. Additionally, orientation can affect detail surface, strength, needed support, and plastic used. We will explore this further in Chapter 12.

**MAKER'S NOTE**

## Set Up Print Parameters

There are a few types of parameters that will be used.

First, set your slicer machine profile to your printer's brand and make; these typically do not change from print to print. Unless you have multiple printers, change printers, or modify your printer, you will set this once and leave it. If you do not have the exact profile you will have to create one manually.

A manual machine profile may include:

- Printer's print volume
- Orientation and shape of the print bed
- Whether or not there is a heated print bed
- Number of extruders
- Specific settings for those extruders – offset, nozzle size
- And any specific G-code variants or printer firmware settings

Next are the material settings. Often slicers have generic profiles for some plastics.

These can be generic settings for each material or specific settings for the material and manufacturer and even color/lot of the material. Often, the slicer can save filament profiles for later use.

The material settings may include:

- Brand
- Material type
- Color
- Density
- Diameter
- Cost
- Weight

For more information, see Chapter 11.

Last are the actual slicing parameters and slicing features. At the most basic level, these should include the following. (For more information on these settings, see Chapter 7.)

- Layer height
- Fill density
- Supports
- Platform adhesion – skirt/brim, raft
- Shell thickness

**MAKER'S NOTE**

There can be over 100 advanced settings, enabling you to control every aspect of the slice and print. The most common ones are included for reference.

Additionally, there are more advanced controls that most slicers allow you to reveal when wanted.

Some of the more commonly used or modified advanced settings are:

- **Shell/wall thickness** (if not in the basic list)
- **Direct control of the number of layers or overall thickness** of the very top and bottom of the object (similar to shell thickness)
- **Extended infill settings**, which include various patterns, overall percentage, how much it overlaps with the outer shell, independent print speeds and angles
- **Extended support settings** which include support density, dense support features, automatic placement, and separation from part settings
- **Fine control of temperature** of the hot end and bed, sometimes per layer
- **Fine control of cooling fan** (sometimes per layer), its top speed, or different speeds for some features
- **Speed settings**, such as default printing speeds, shell/wall speeds, infill speeds, support structure speeds, and general non-printing movement speed
- **First layer settings** (like height, width, speed) are often different to help with adhering to print bed
- **Bridging settings** control how the printer handles printing a span of filament between edges, allowing printing of short spans with little or no support and with minimum quality loss
- **Ooze control** (ways to mitigate material oozing from the nozzle when moving), which includes retraction distance, retraction speed, coasting distance, wipe distance, z-hop (or vertical lift)
- **Prime pillar and ooze shield** are slicer-dependent features useful for multi-color prints

Some slicers have special modes or experimental settings such as:

- **Vase mode** prints in one continuous filament, slowly spiraling upwards. This is limited to basic shapes with no infill and one shell/wall.
- **Draft shield** makes a wall around the object being printed to cocoon the object and keep it isolated from the environment. It is best used with materials like ABS that can warp or split if cool air causes uneven cooling and the printer does not have a full or heated enclosure.
- **Multiple setting profiles or dynamic settings** that can change during print. For example, a change in layer height depending on curve of ob-

ject, a change in infill to start low and increase near the top of the object to support it, and so on.

- **"Fuzzy" exterior** can add some random noise to the exterior shell to create a rough texture.
- **Ability to print multiple objects on the print bed one at a time** rather than all one layer at a time. This limits you in the number of objects printed but allows a print to be stopped and objects finished and removed before the entire print is completed or fails.

## Slicing your model

This takes a 3-dimensional manifold (watertight) object and cuts it into slices, dictated by your layer height. These slices are turned into paths and saved as G-code (or similar) that a 3D printer can read and follow.

One good way to think of slicing is like a mandolin or a deli meat slicer! The object is sliced into cross-sections of the thickness you want. The thinner they are, the more numerous they are, and the more time it takes and the finer detail each cross-section can hold.

After the slicer slices the object, the slices are defined by paths (think of tracing your finger). These paths are generated into G-code that describes each layer. G-code is a computer language that communicates the slicing information to the printer. This includes speed, location, path directions, and filament flow, to name a few.

Most slicers will render a representation of the sliced object. You can even step through the slices and view each layer or simulate the entire printing process.

At this point, the G-code can be copied or streamed to your printer and it will follow the instructions, outlining each layer one at a time on top of the next, building your object.

# › 9.2 – Do I have to worry about safety and ventilation?

*If you are using a print service, you can skip this section.*

When you have your own printer workspace, there are always safety concerns, as nothing is truly 100% safe. For 3D printing, some potentially dangerous elements to understand are the temperature of the build platform and print head, the mechanical movement of the printer itself, and that the printer is an electronic device. Additional concerns include the molten plastic and any particles or gasses the plastic exudes during the process, as well as any post-processing you do.

Burns, electrical shock, fire, and fumes are possible. Follow some simple safety protocols:

- Follow all manufacturers' instructions and safety instructions.
- Always Inspect the printer and verify it is in good working order.
- Power the printer correctly. The heaters take some power to operate, so use correctly rated extension cords, a plug that can handle the draw, keep it dry, and so on.
- Keep your hands away from the inside of the printer when it is running.
- Don't touch heated surfaces with bare fingers.
- Know what you are printing and its properties!
- Ventilate accordingly.

A great rule of thumb is do more than you need.

Ventilation is a big concern with a wide range of responses for how and how much you should do. A common rule of thumb is: if you can smell it, you may need to have more ventilation. This is not a very accurate metric, as you cannot smell all of the chemicals being emitted. Assume a base amount and study the material data sheets closely.

For example, people assume PLA is harmless because it is "organic" and smells good. This is not the case; how it smells or how "organic" it is is meaningless for health and safety. Keep in mind that all petrochemicals are organic molecules; it is chemical composition coupled with dose that makes something toxic.

In short, it is safest to ventilate. Given the variability in materials and even formulas within those materials, it is best to be safe.

## Filters

Filters work in the best cases, but often they can't supplant or compete with simple ventilation. The right filter may be able to absorb the particulate matter, but only the best will also absorb gasses in the amounts needed.

If you use filters, it is crucial that you replace them regularly! Overusing a filter is as bad as not having one.

## › 9.3 – Why do I need to run hardware tests?

*If you are using a print service, you can skip this section.*

When you buy or build your own 3D printer, you must calibrate and test your settings.

When you are printing for your first time with a new printer, it is important to get a baseline print to check out and verify the printer's functionality (i.e., no shipping damage, loose parts, functioning sensors, etc.).

1. Make sure the printer is unpacked and assembled according to the directions.
2. Double-check those instructions. You might have missed a step – like tape or shipping support – and then spot-check that parts move correctly and have no obvious damage.
3. Place the printer on a level surface in a room with ventilation.
4. Power it on and check if it starts correctly. If needed, consult your manuals.
5. Install the software that comes with the printer. Normally this is a preferred or proprietary slicer with a custom printer profile for your printer.

   If the printer has controls onboard, some of these steps can be done locally on the machine, otherwise perform these steps with the software either attached with a USB cable or wireless.
6. You may want to level your print bed.

   Note that some printers automatically level – for these you can just enable that function.

   If you level it manually, follow your printers' instructions. Generally, this consists of moving the print head to the home position (one corner and bed just touching the print head). Then either follow an automated procedure of the head moving to different points (or manual movement – with the stepper motors off) and moving the screws under the print bed. This is a very important part.

## General manual bed leveling/tramming and gap instructions

A good rule of thumb distance is ~0.2 mm. It is about the thickness of a piece of paper. A good piece of paper to use is a standard 8.5 x 11 (letter-size) piece of copier/printer paper. For easy handling, cut it into index card-sized pieces. This helps to make the gap as uniform as possible and creates just enough of a gap to allow the plastic to make a bead.

Manual printers will have bolts or knobs with springs that control the movement of the print bed.

You should first "Home your printer" – this should be done in software or firmware. This is effectively the default starting position of the printer. The print head should be almost touching the print platform.

Ideally, use the leveling or tramming function built into the software or firmware. This normally moves the print head to 4–5 locations on the print platform on command and steps you though the process.

If this function is not part of the software or firmware, turn your printer's stepper motors off (or turn your printer off) to allow you to manually move the print head.

Now place a single sheet of the paper you cut earlier under the print head. It should slide under; if it doesn't, use the bed leveling/tramming knobs to lower the bed. Once that is done, move the bed leveling/tramming knobs slowly while moving the paper back and forth. When you start feeling resistance on moving the paper, stop. The head is pinching the paper between itself and the bed. You should be able to move the paper but feel a bit of resistance. Now move the print head and repeat. Normally I like adjusting all the corners, then the middle, and then start over again to ensure all are calibrated.

This should both tram your print bed and ensure the bed maintains the same distance between the bed and the print head throughout its traverse over the bed (see Figures 9.2–9.5).

To test your print bed level and gap, there are a few very simple objects available from various sites that are only a few layers thick and cover most of the print bed to test uniformity. They are generally a large "X" or

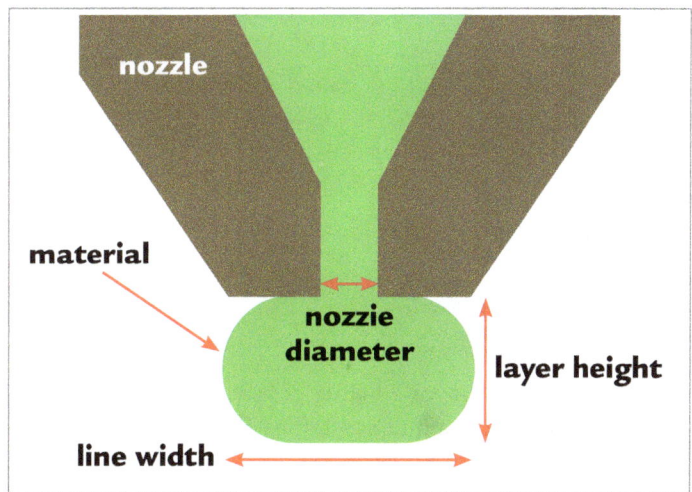

FIGURE 9.2 – This is an illustration of the print nozzle with the material exiting it, making a bead. Note the nozzle diameter is less than the line flat and total width, as the bead needs to be squashed to stick to the bead and every layer.

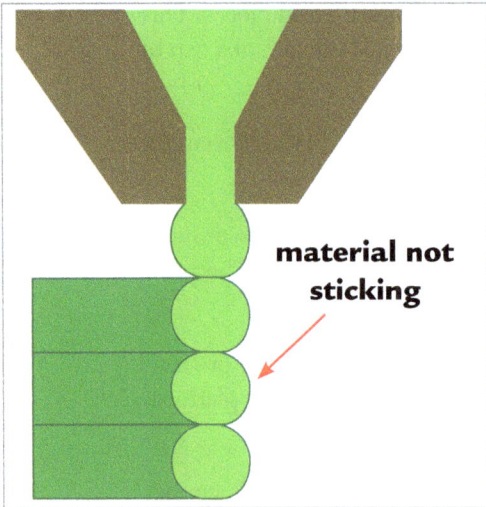

FIGURE 9.3 – If the head is too high, the plastic will not stick to the bed or other layers.

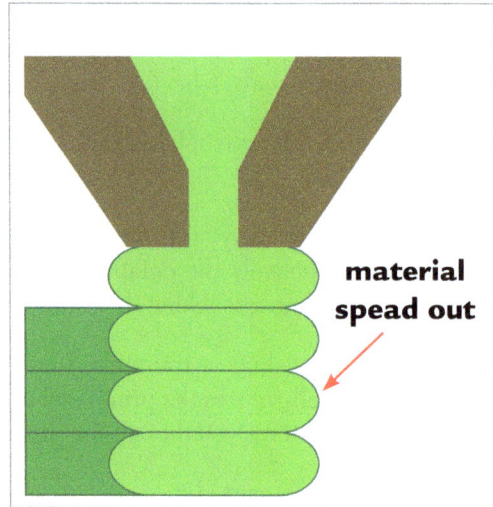

FIGURE 9.4 – Too low and it may not come out of the nozzle at all, potentially jamming the print head or stripping the filament, or causing it to be too spread out.

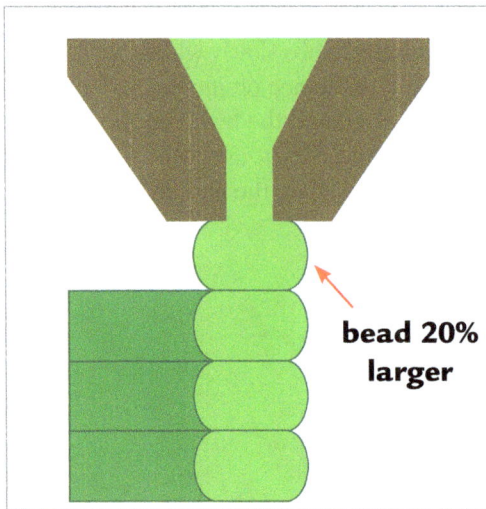

FIGURE 9.5 – Ideally, you need a bead of plastic that is around 20% larger than the nozzle.

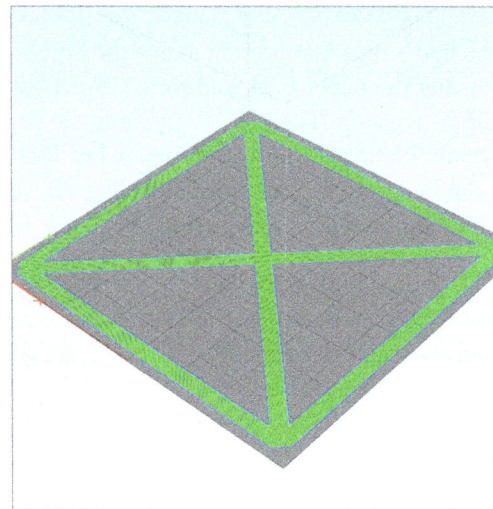

FIGURE 9.6 – A bed leveling object of just two layers, covering most of the build plate. Jonathan Torta.

"O" shape or a matrix of squares across the build plate. These can be handy to carry out a quick test. (See **Figure 9.6**.)

Note that no print bed is perfectly flat; some warp slightly under stress or heat. Your test may not be perfect over 100% of the print plate. Some firmware and auto leveling/tramming settings can take this into account.

Otherwise, calibrate your print bed as accurately as you can and focus on the space you will use most frequently, like the middle of the print bed. If you must, you can always print with a raft to ensure an even platform and good bed adhesion.

A bad leveling or tramming job or gap can result in failed prints, so this is important to set correctly and to check periodically.

Newer printers with auto leveling/tramming features are very nice, but still may need fine manual or software tweaks. Note that some firmware or software can modify the overall gap with a Z-height setting. This allows subtle changes in the Z-height that will change the first layer gap.

## Indicators of poor leveling/tramming and gap

If the gap is too big or too small, you can often tell by analyzing the quality of the extruded lines of filament on the print bed. (See **Figure 9.7**.)

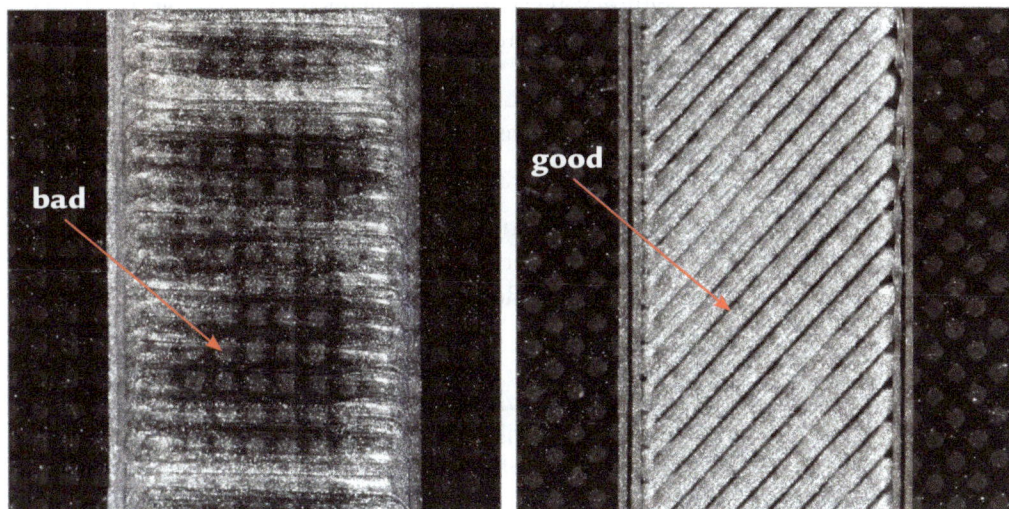

FIGURE 9.7 – **A.** A bad result. The top is too close to the build plate; the filament is not sticking and eventually stops flowing. **B.** A good result. Nice and even. Jonathan Torta.

- **The first layer is barely visible or very thin:** This indicates the nozzle gap is too small and restricting plastic flow. Increase the gap.
- **No plastic extruded onto the print bed:** The nozzle is so close to the print bed that it simply cannot extrude the filament. Increase the gap.
- **Plastic gathers on the nozzle:** The nozzle is too close to the build plate and is picking up plastic or the filament is not sticking and curling. You may need to watch the print in progress to see which is happening.

- **Extruded filament doesn't stick to the build surface:** The nozzle needs to physically press down on each extruded line of filament enough for it to squash onto the print bed and stick. If the nozzle is too far away, it will not stick and will move when the print head comes around again. Decrease the gap.
- **Extruded filament comes out as spaghetti:** A severe case of the nozzle being too high above the build plate and effectively extruding into the air. Decrease the gap.

## Load Your Filament

Set the printer to warm up. Many printers have a cleaning, preheat, or fill filament loading you can use.

Monitor the temperature and make sure the print head and the build platform heat up to the temperature set (often ~200°C print head for PLA). After it reaches this temperature, you can load your filament. Follow your printer's instructions, as each printer is different.

We would suggest that you use PLA for the first print, as it is the easiest to print with and thus easier to create a baseline, because you introduce fewer new filament variables.

After the filament is loaded, extrude some and make sure it flows out of the nozzle in a nice thin line. Look for these features in your instructions; often the loading procedure includes this. Remove any excess filament and let the printer cool down.

**MAKER'S NOTE**

Do not let PLA sit in the nozzle for extended periods of time at printing temperature as it will start to decompose and may clog your nozzle (i.e., at print temp without extruding for over 10–15 min – if you do, be sure to extrude that filament to get to fresh plastic before printing.)

You are now ready to start your first print. This first print should be simple to just test the functionality of the printer and all its parts.

1. Load the object.

   Select a very simple object like a small cube (included in the extras and on DVD) into your slicer.

   Typically, this will drop your cube in the middle of the print space on one flat side. If not, follow the instructions in the slicer and do so.

2. Use the defaults in the slicer and make sure your printer's profile is selected and the correct filament type is also selected.

   If the slicer requests a layer height, select the lowest resolution, low or ~0.3 mm. There is no need for a skirt or support.

3. Slice the object in the program – often this will give you a preview of how it looks sliced.

4. Accept this and save the sliced file and/or and initiate the print and let it run.

5. Watch the printer warm up again and follow its progress, make sure of its correct operation, that all parts are moving correctly (no binding or hitting the limits), and if anything goes wrong, turn it off.

Most common issues are first layer issues, the print not sticking, the head being too close, or the bed is not level.

- Make sure the print bed is fully clean. Use alcohol to clean it off, as any oils can hinder adhesion.
- PLA can use up to 60°C of heat; if you have a heated bed, this can help the initial layer stick. But it often is not needed and just uses energy. Also, people have used other aids such as glue stick (an Acrylic polymer) or hair spray to aid adhesion.
- If the print does not stick, recheck the bed level and gap. The first layer should be printed slowly and squashed onto the print bed.
  › If beaded up too high (because of too much of a gap), it may just pop off.
  › If it's too close, no filament can escape (or worse, it scrapes on the print bed). This can make the stepper motor skip or the feed gear dig into the filament, stopping any further feeding.
- If the layers are misaligned or shifted, there may be something wrong with the printer's movement. Check for packing leftovers, bent rods, loose screws, and other mechanical problems.

Ideally you will get a print of the cube; it may not be perfect, but we are looking for basic printer functionality at this point.

- If not, recheck the basic settings and filament recommendations.
- If the printer has serious problems, contact your printer's service team.
- If the print looks more or less good, we can move on to calibration.

## › 9.4 – How can I transfer my file to my printer?

There are a few ways to transfer files to the printer. The most basic printers use a SD card that contains the sliced object (normally the G-code); you copy the file from your computer and then walk the SD card over to your printer, insert it, and run the print from the printer.

Alternatively, you can connect to the printer directly with a USB cable and feed the information directly. This method is also how you can directly control and update firmware. This method, while convenient, can be an issue if the file is streamed to the printer, as the connected computer has to remain on and can't be interrupted or the print will fail. This is not an issue when the file is copied to the printer in its entirety first.

Newer printers are usually Wi-Fi enabled and can copy or stream. This method has the same issues if the computer stream fails or if there is a problem with the Wi-Fi.

Alternately, a dedicated secondary computer or a printer with more than just the basic controller can do all of this. Astroprint/Octoprint is a dedicated Raspberry Pi microcomputer that you can copy the G-code to (with Wi-Fi) and even directly slice with and then stream the data to the printer. These devices are very flexible and even allow you to remotely control and monitor your printer over Wi-Fi or a mobile app. Some high-end printers have an equivalent built-in feature (see **Figure 9.8**).

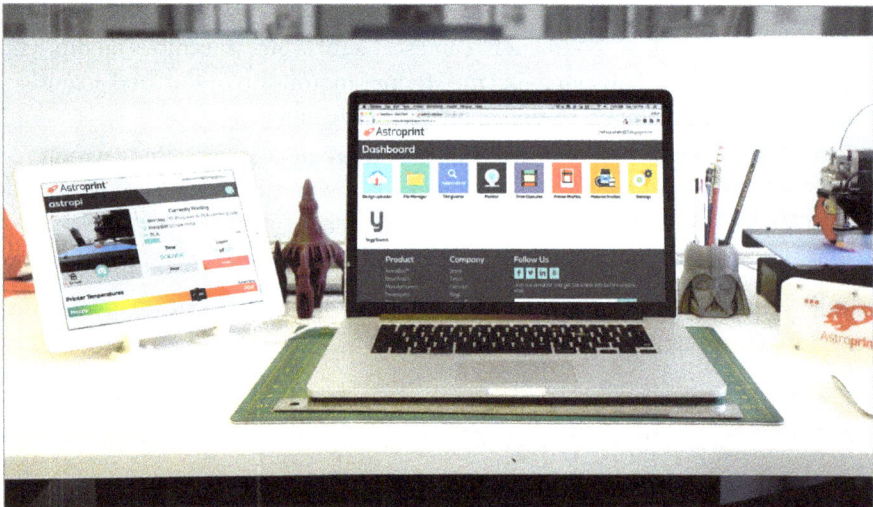

FIGURE 9.8 – AstroPrint Cloud interface when managing a 3D Printer wirelessly. JayLoerns.

## › 9.5 – What happens during a print? Is there anything I need to watch out for?

Pay special attention when monitoring the first layer of the print. Levelling or tramming the bed is very important; the first layer's success and quality can determine the success of the rest of the print.

- When the print head is too far from the bed, the object can become unstuck; if it completely detaches, the printer will start printing in the air – making what we like to call a "birds' nest" – or worse, the filament gets stuck under the print head and makes a massive ball of plastic, potentially damaging the printer (see **Figure 9.9**).
- When the head is too close, the head may crash into the bed and damage it, or more often, the filament can no longer exit the nozzle. This can jam the print head; the stepper motor will skip or strip the filament.
- When the print head is in the good zone, you will have an even layer that will stick nicely for the entire print.

Next, make sure you have sufficient filament on the spool to complete the print. Some printers allow a pause for a filament change and restart; others have a sensor that automatically pauses the printer and gives a signal. See Figures **9.9–9.11** for sample images of errors.

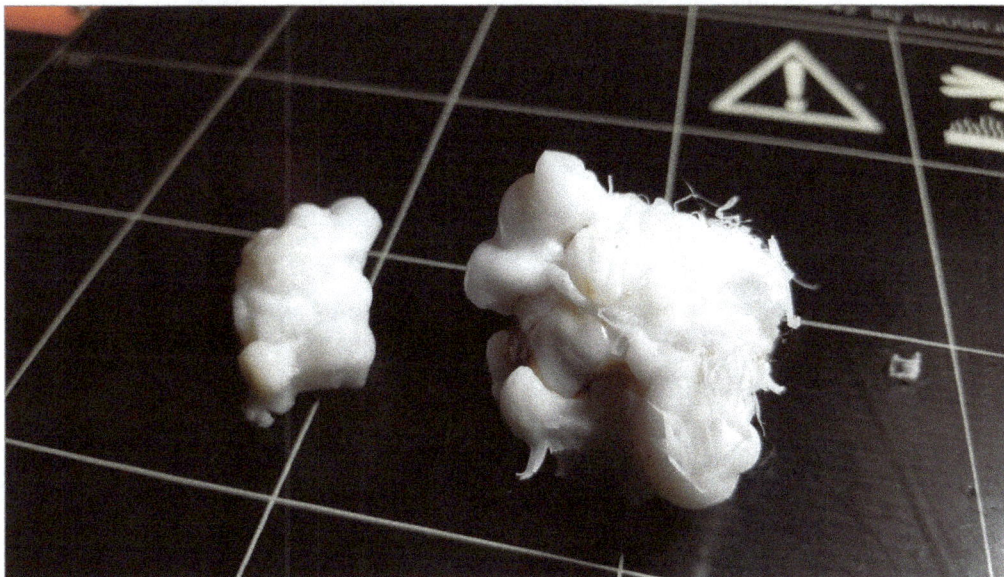

FIGURE 9.9 – Blob of filament. Lucas Phillips.

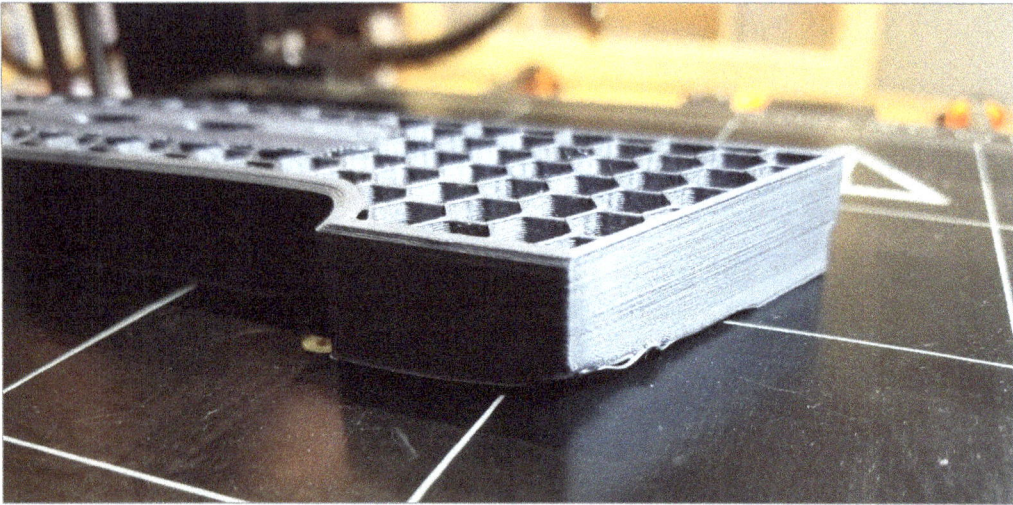

FIGURE 9.10 – Curling of print edge. Lucas Phillips.

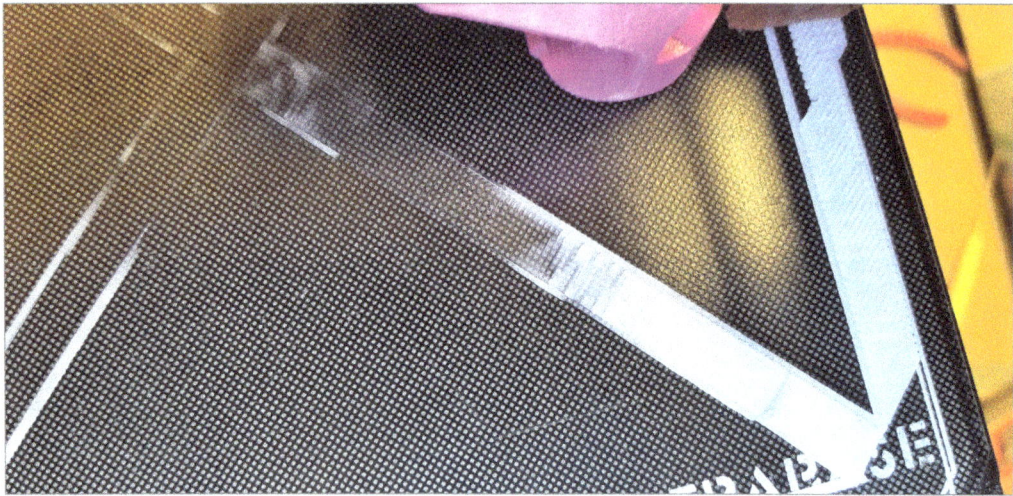

FIGURE 9.11 – Bad first layer. Jonathan Torta.

## › 9.6 – Are there ways to remotely monitor my printer?

There are a number of applications for monitoring your 3D printer remotely. Both Octoprint and Astroprint top the list for full-featured control, including remote by browser or by app (phone or tablet). Other programs like Repetier-Server allow full DIY control, or more straightforward control like waggle.

Some printers have these functions built-in using their own services.

https://creatablelabs.com/waggle/     https://www.repetier-server.com

Web link

FIGURE 9.12 – Duper Pro 3D printer with remote monitoring. Thelema4ever.

## Extra Additions Including Wi-Fi, Cameras, and Monitoring

Note that each product uses its own method. You may need to open a port on your router. If you do not open some security for it, you may lose some functionality (see **Figure 9.12**).

## SUMMARY

Setting up to print is vital for making a 3D object. After your model is verified as acceptable, you can start loading your slicer application. There are different types of parameters that can be used. There are also different methods of transferring your files to your printer. You will have to investigate which is best for you. Safety and ventilation are your top priorities to prevent burns, electrical shock, fire, and fumes. It is important to get a baseline print to check out functionality of the printer. A bad leveling/tramming job or gap can result in failed prints, so this is important to set correctly and check periodically. When printing, you need to watch it start and lay down the first layer, because this will determine the success and quality of the rest of the print. There are many applications that will monitor your 3D printer remotely if needed.

## APPLYING WHAT YOU'VE LEARNED

1. Continue making your own 3D dictionary by adding the definition (in your own words) using five words related to 3D printing in this chapter.
2. Explain why orientation is very important when using a 3D printer.

3. Name and explain three types of parameters that can be used.
4. Pick ten different advanced settings of slicing and printing and explain why you would use them.
5. What are three ways you can transfer your files to the 3D printer? Explain the advantages of each.
6. Why is it important to monitor what happens when your object is printing and how can you do it?
7. Explain why safety and ventilation should be your top priority.
8. Why do you need to get a baseline print?
9. What are the signs of a bad leveling/tramming job or gap and what can you do about it?
10. How do you clean a printer bed?

# PART **2**

# Using Your 3D Printer

Part 2 helps with initial calibration of the printer and the filament, as well as a few objects to help with solving common issues with printing. Calibration is often overlooked and skipped because people just want to start printing. This often leads to problems and disillusionment as the first prints are at best fair, and at worst a disaster. Additionally, having in mind the limitations and workarounds with 3D printing can further enhance the quality and widen the range of possible prints.

The goal of these chapters is for you to see through the eye of the maker by not only enumerating the printing steps but also adding a running commentary from the maker. We'll include firsthand tips and explanations highlighting possible pitfalls to watch out for and best practices.

Each challenge was created to teach selected skills and showcase different challenges that build upon each other. We have also included project files on the DVD so you can print your own projects.

By Creative Tools

# Calibration

## OVERVIEW AND LEARNING OBJECTIVES

**In this chapter:**

- 10.1 – What is the importance of calibrating my printer?
- 10.2 – The calibration project Mow CAL
- 10.3 – Where can I go for troubleshooting help?

## › 10.1 – What is the importance of calibrating my printer?

Using a 3D printer can be complex, with a lot of variables. An uncalibrated printer can give poor or unexpected results. A quick calibration test and review can fix minor issues very quickly. More in-depth calibration can fine-tune your printer. Also, each plastic type brand and formulation will print slightly differently on each printer. So, some sort of calibration to **your** setup is required, as default settings are only ballpark.

Do yourself a favor and record the settings for your printer and any calibration changes you make. This will give you an understanding of your own printer. You will have quick reference for what each change does, how much of a change does what, and if your printer is particularly good or bad at this or that function, along with potential mods that can correct that limitation.

**MAKER'S NOTE**

For example, one of my printers had a rather poor filament cooling fan and duct (a known limitation). This was identified, and I found an additional issue, which was it blew air only on one side, making the limited air even worse at cooling the filament.

A quick search on Thingiverse allowed me to find a better duct and I purchased a better fan (more cfm). This worked around the limitation and allowed faster printing, better spanning, and overall better printing with PLA. But I decided to make my own, and through collaboration with others came up with a highly enhanced cooling duct that allowed the air to hit from all angles and mount a much more powerful fan. This two-stage process solved my cooling issues.

Chapter 11 contains additional information on how each plastic functions with different settings, with recommendations on how to record your settings for each plastic you print with.

## › 10.2 – The calibration project Mow CAL

For the first calibration project, we created a complex calibration test object to fine-tune your printer and make sure everything is running well: the "Mow CAL." (See **Figure 10.1**.)

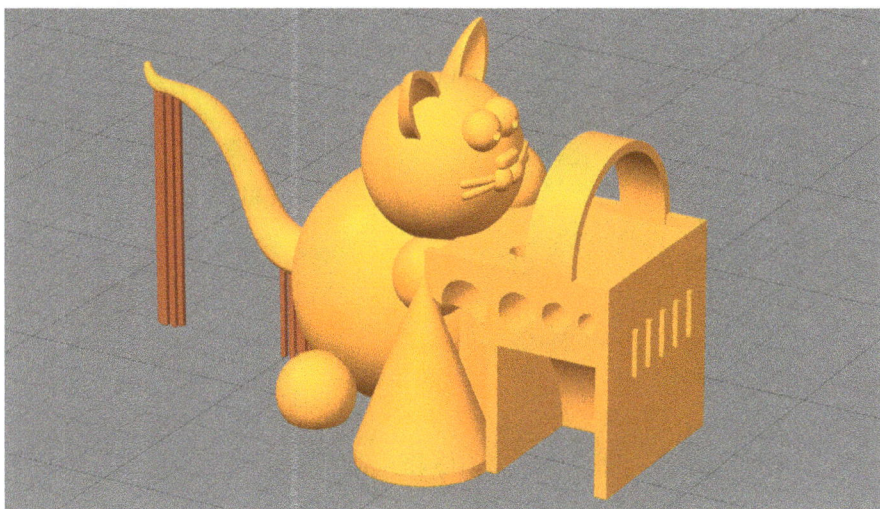

FIGURE 10.1 – The "Mow CAL" test object. Jonathan Torta.

*Project files along with color images and videos are included in the extras and on the DVD.*

IN EXTRAS

This calibration model includes boxes, spheres, and cones, raised and embedded text, holes, curves, arches, overhangs, hollow objects, various angles, and detail.

## Mow CAL features and what they do

**Shapes.** Each shape helps check the surface quality of the print. A flat surface checks for smoothness and even ghosting; other shapes check how the printer handles more than one dimension or larger angles.

For example, a cone checks the evenness of round layers at different diameters. The included cone can also check for stringing with the rest of the object. The spheres check round layers of more than one diameter and layer angle (overhang and stepping). (See **Figures 10.2 and 10.4**.)

**The horizontal holes** check for dimensional correctness in shape – for example, roundness – and also size: 1 mm, 1.5 mm, 2 mm, and 2.5 mm diameters. Note that these have a fillet (a round bevel) on one side and a hard edge on the other (see **Figures 10.2–10.3**).

The top of the box has **3 different thickness edges:** 0.4 mm, 0.8 mm, and 1 mm. It is important to note that the 0.4 mm wall may or may not print, depending on your slicer and settings, as a nozzle of 0.4 or larger may not be able to print this to

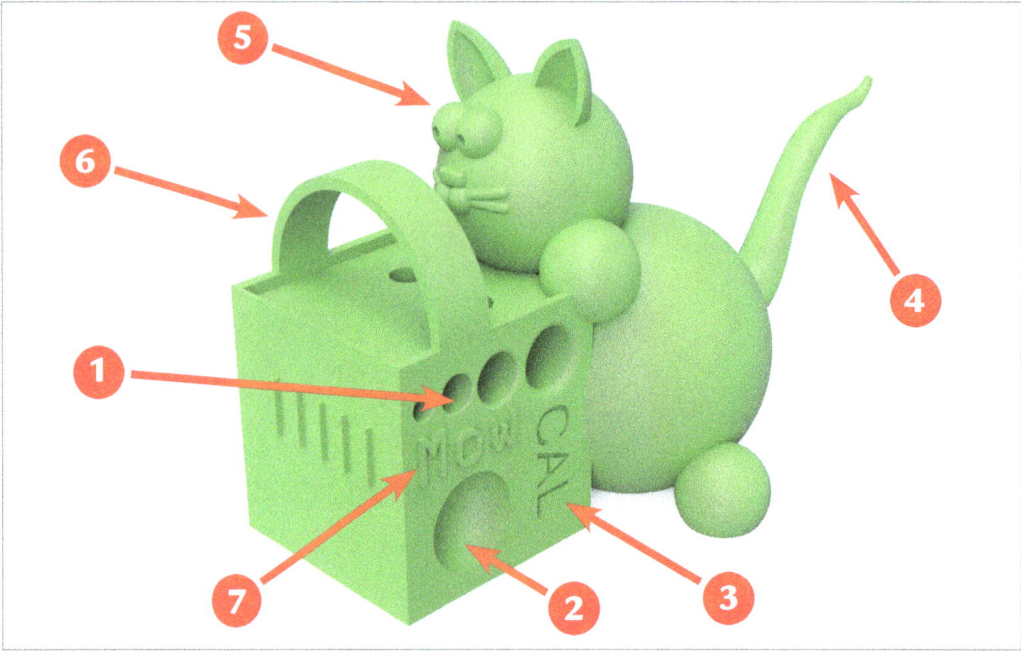

FIGURE 10.2 – 1. Holes (1-1.5-2.2.5mm Filleted), 2. Impression (surface quality), 3. Embossed text (small detail), 4. Tail (overhang), 5. Face Details (surface quality), 6. Arch (overhang), 7. Extruded text (small detail). Jonathan Torta.

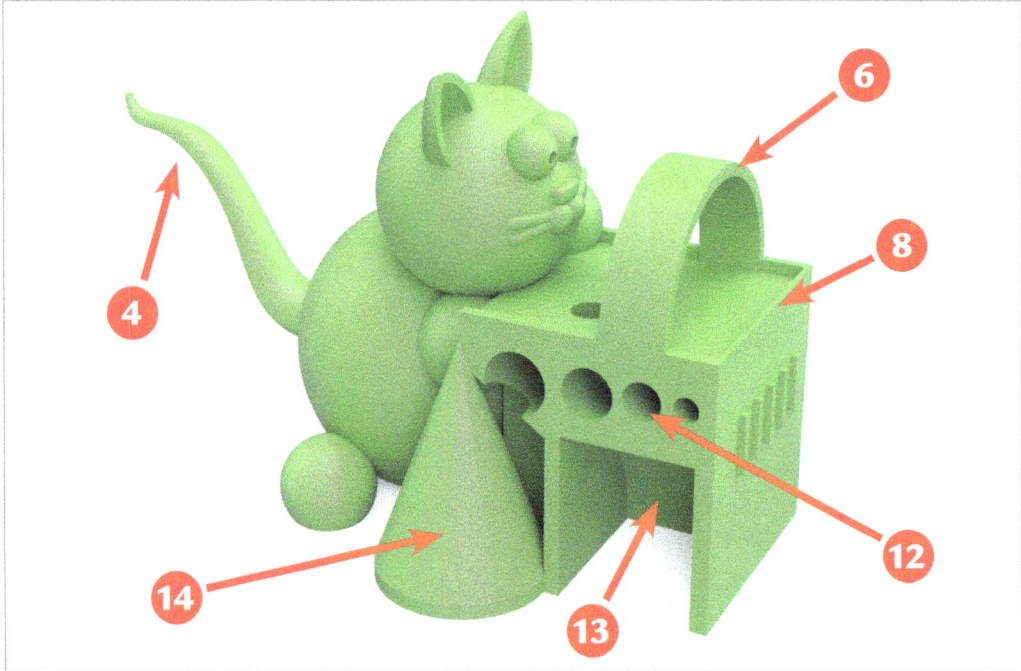

FIGURE 10.3 – 4. Tail (overhang), 6. Arch (overhang), 8. Various width edges (small detail), 12. Holes (1-1.5-2.2.5mm Hard edge), 13. Arch (spanning), 14. Cone (surface quality). Jonathan Torta.

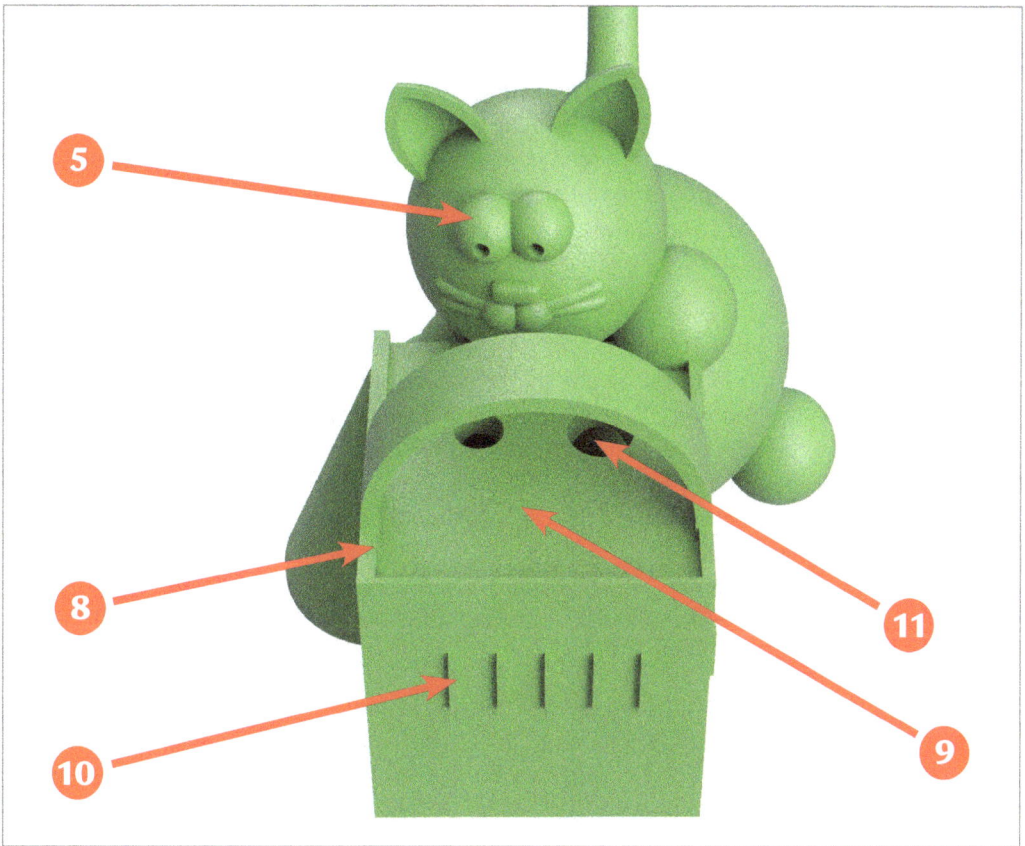

FIGURE 10.4 – 5. Face details (surface quality), 8. Various width edges (small detail), 9. Flat top (smoothness), 10. Bars (Ringing), 11. Holes top (3-4mm). Jonathan Torta.

size. The 0.8 mm wall could be one or two wall thicknesses, and the 1 mm is 2.5 walls (depending on your slicer settings). How your slicer handles these is important to note. For example, the 2.5 mm wall may have a gap between two walls (see **Figure 10.4**).

Other details are the **embossed and extruded text**, as well as detail in the cat's face. These can all produce ghosting (vibrations of change of direction). (See **Figure 10.2**.)

The **arch** on the side of the box shows spanning at different angles. (See **Figure 10.3**.)

**The top arch** can show cooling issues and overhang ability of the filament. (See **Figures 10.2–10.3**.)

**The tail** can show how an unsupported detail fares. (See **Figure 10.2**.)

Bonus items are that only some areas have infill. You can check the tops of these for sagging; you might need more infill. The main body of the cat is hollow (with a hole on bottom) to check how the unsupported shell prints at various layer angles.

Printing this calibration model perfectly is a very difficult task and may not be possible for all printers and materials. This is fine; it was our design choice. Seeing how your printer handles each feature is as important as fine-tuning it.

For instance, if you have a lot of ghosting, you may need to tighten up your printer (loose screws) or put the printer on a flexible surface or make flexible feet to deaden vibration. All printers will show some vibration at high speeds; printers with heavier print heads potentially suffer more. Overall rigidity of the printer plays a big factor and braces can help. Just lowering the speed will reduce the effect. Changes in your printer's firmware in relation to acceleration and jerk may help.

As you see, there are many changes and fixes (minor and major) that can be used to help with the issue.

Keep in mind that "good enough" is your goal, because perfect is not possible. Additionally, every modification and tweak can affect other aspects. Do them one at a time and recheck, as this allows you to identify and back out a change.

## Equipment and materials used

It is best to start with PLA and your default PLA profile for your printer and change from there.

Also, be sure you have followed the first-time print recommendations in the previous chapters to ensure the printer is basically functioning and has no parts damaged by shipping.

This also includes leveling the print bed for printers that require manual leveling. Remember that a level bed is very important and can be the difference between a print failure and success (see **Figure 10.5**).

## Printing the Mow CAL

To print the Mow CAL:

1.  Load the file into your slicer and make sure it is on the print bed and centered.

## CORRECT LAYER HIGHT

bead

nozzle

material

print bed

TOO HIGH

TOO LOW

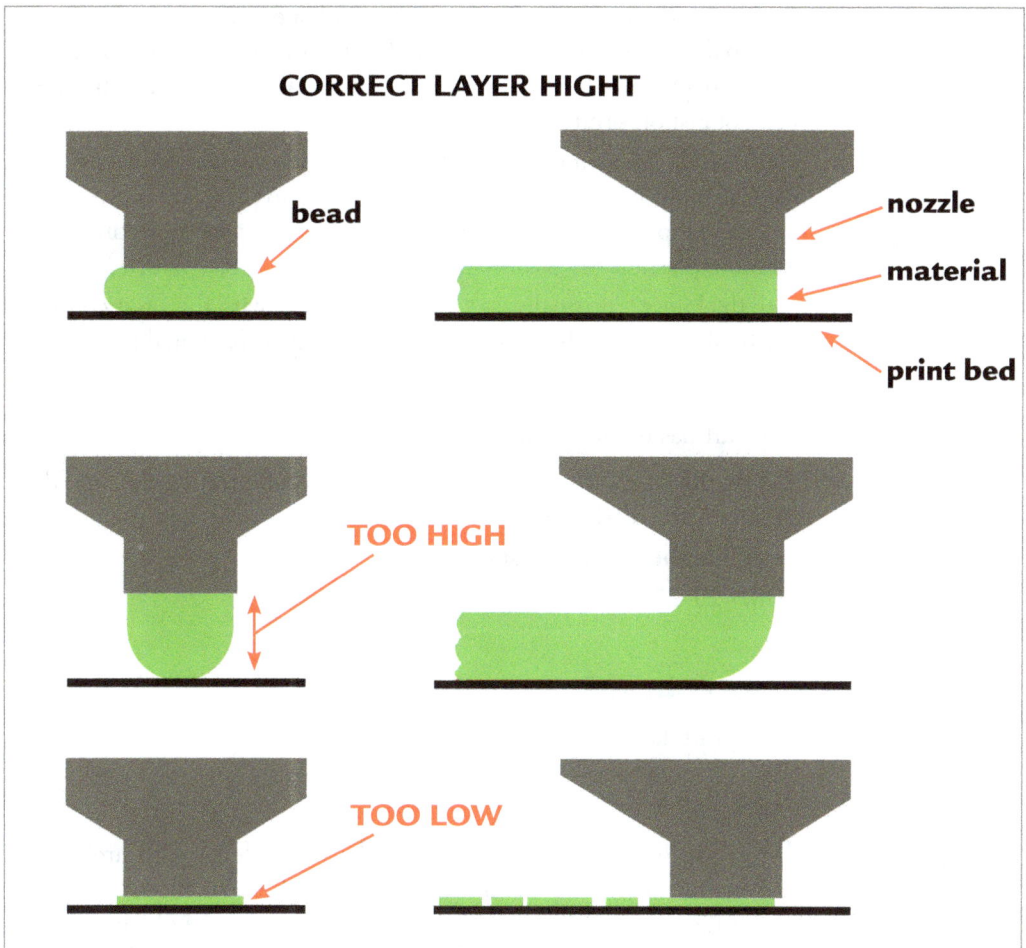

FIGURE 10.5 – Initial layer hight is extremely important too high and little surface area touching the bed and it will not stick, too low and the material may not flow consistently or at all, just right is a mildly swashed bead of material.

2. Select a profile. For example, match your plastic settings to the material you are using.

3. Select a layer height (start thick and progress lower). Suggested starting layer heights are 0.3 mm or 0.2 mm.

4. This object does not *need* supports, but you can use them to see how they function.

   Note that under the tail is from bed and it changes diameter – some slicers may not support it under its entire length automatically. This forces you to change parameters or manually put in support if needed – it is good to know the minimum detail size of your slicer or how to change it.

5. There is a support under the two arches and one under the cube. This is good to determine and set the angle the supports will start supporting. The support under the top arch shows how enabling supports anywhere (i.e., not just off print bed) will work.

6. Set the infill. You can experiment with this – a lower infill will save material but may make some top layers harder to print smoothly. More infill can help the quality of top of the box but also take more time and use more plastic.

7. Shell thickness is normally 2–3. Feel free to play with this, as it contributes to object strength but also could help with surface quality if there are more than two shells.

A good first start has the following parameters:

- **PLA profile:** for most printers, this sets the temperatures, fan, and the potentially the feed rate.
- **Layer height:** 0.3 mm to start, then 0.2 mm
- **Infill:** ~20–25%

## Analyzing your print

Analyze your print, keeping in mind some quality metrics.

### Things to Look For

How did it print? Is it clean overall? Did it have problems with arches and overhangs? Are the surfaces clean and smooth? Are there small holes or blobs of plastic over the surface? Are there wispy strings all over the object? Are the holes round? Can you read the text? How did the bottom of the tail print? Did all of the thin walls on the top of the box print?

### Common Problems

Take a good look at your print and try to match the issues. Only work on one issue at a time, as most settings affect others and modifying more than one can give unpredictable results.

Note that this outline is just that – a brief list of possible solutions. The ultimate solution can be quite complicated or be tied to your specific printer, filament, or setup.

See the links at the end of this section for references to various problem lists.

**Not extruding** – Printer does not extrude plastic at the beginning of the print.

- Extruder is clogged
- Filament is stripped
- Filament is jammed
- Heat creep for PLA

**Not extruding at start of print** – Print grooves left in bed with no filament and/or intermittent lines of filament on the first layer.

- Extruder not primed before printing object
- Nozzle too close to print bed
- Print bed is not level

**Not sticking to bed** – The first layer does not stick to the bed and the print quickly fails

- Build platform is not level
- Filament not pressed against the print bed well enough. (The nozzle may be too high?)
- First layer printed too fast
- Filament or bed temp too low. Filament cooling fan on for first layer
- The build platform condition – cleaning, material, supplement stick materials, problematic print material
- Too little surface area for good adhesion.

**Under-extrusion** – Printer does not extrude enough plastic, gaps between perimeters and infill

- Filament diameter / nozzle setting incorrect.
- Extrusion multiplier setting incorrect.
- Partially clogged nozzle

**Over-extrusion** – Printer extrudes too much plastic, prints looks very messy

- Filament diameter / nozzle setting incorrect.
- Extrusion multiplier setting incorrect.

**Gaps in top layers** – Holes or gaps in the top layers of the print

- Under extrusion
- Need additional top layers/thickness.
- Low infill percentage (excessive internal bridging)
- Too much cooling

**Stringing or oozing** – Strings and hairs left behind when moving between different sections of the print

- Retraction distance / speed settings. For direct drive, usually 0.5 to 2.5 mm. For Bowden extruders, 5.0 to 8.5 mm seems to work well.
- High filament temperature
- Combination of movement in open spaces (without printing) and travel movement speed

**Overheating** – Small features become overheated and deformed (See **Figure 10.6**)

- Insufficient filament cooling.
- Environment (hot environment, enclosed print space)
- Printing too little, too fast – filament can't cool fast enough

**Layer shifting** – Layers are misaligned and shift relative to one another

- Mechanical problems (slipping belts, movement blocked, stepper motors missing steps)

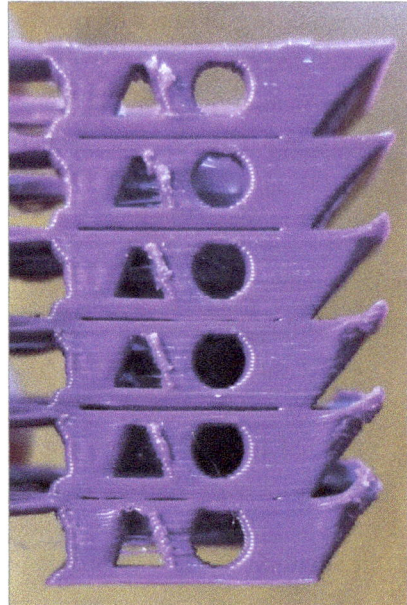

FIGURE 10.6 – Small features become overheated and deformed. Jonathan Torta.

**Layer separation and splitting** – Layers are separating and splitting apart while printing

- Print temperature too low
- Too much cooling
- Layer height too large

**Grinding filament** – Plastic is being ground away until the filament no longer moves, "stripped" filament

- Filament movement too fast (retraction, or extrusion)
- Filament too cold (not melting fast enough)
- Nozzle clog

**Stops extruding mid-print** – Printer stops extruding plastic randomly in the middle of a print

- Not extruding

- Power or printer issue (restart)
- Issue with G-code file or transmission
- Heat creep for PLA
- Out of filament

**Weak/stringy infill** – Thin, stringy infill that creates a weak interior and does not bond well

- Lower infill print speed
- Infill extrusion width too low

**Blobs and zits** – Small blobs on the surface of print, otherwise known as zits

- Incorrect retraction and coasting settings

**Gaps between infill and outline** – Gaps between the outline of the part and the outer solid infill layers

- Need more shell overlap
- Print speed too high
- Acceleration/deceleration issues (lash and loose parts)

**Curling or rough corners** – Corners of the print tend to curl and deform after they are printed

- Overheating

**Fat first layer/elephant's foot** – Bottom layer is larger and bulges out from the sides

- First layer's height is too low
- Level bed
- Too high of an extrusion rate for first layer

**Scars on top surface** – The nozzle drags across the top of the print and creates a scar on the surface

- Set Z-hop
- Over-extrusion

**Gaps in floor corners** – Gaps in the corners of the print, where the top layer does not join to the outline of the next layer

- Infill percentage is too low
- Add additional shells or top layers

**Lines on the side of print** – Side walls are not smooth, lines are visible on the side of the print

- A variation in temperature (filament cooling fan cooling hot end)
- Inconsistent extrusion
- Mechanical problems

**Vibrations and ringing** – Vibrations that cause oscillations on the surface of the print, otherwise known as "ringing"

- Printing too fast
- Acceleration setting in firmware
- Mechanical problems

**Gaps in thin walls** – Gaps between thin walls of the print where the perimeters do not touch

- Extrusion width
- Thin wall settings

**Small features not printed** – Very small features are not printed or are missing from the software preview

- Single wall settings
- Too thin a feature
- Too wide a nozzle
- Change orientation

**Inconsistent extrusion** – Extrusion amount tends to vary and is not consistent enough to produce an accurate shape

- Filament binding (make sure the filament can move freely)
- Partial clog of the extruder
- Fulfillment quality issues
- Mechanical problems

**Warping** – Warping of large parts, particularly with high temperature materials such as ABS

- Filament cooling on first layers (turn off cooling for these layers)
- Materials like ABS require an enclosed or heated print chamber
- Minimal surface area (use brim, raft or change orientation)

**Poor surface above supports** – Poor surface quality on the underside of the part where it touches the support structures

- Support settings (vertical/horizontal separation and support density)

**Dimensional accuracy** – Dimensional issues where the measured dimensions do not match the original design intent

- First layer (first layer can be of a different size then the rest of the layers)
- Under- or over-extrusion
- Printer issue (one dimension constantly off – use slicer or firmware settings to calibrate)
- Material factors (shrinkage)

**Poor bridging** – Sagging, drooping, or gaps between the extruded segments of your bridging regions (see **Figure 10.7**).

- The slicer notices this is a bridge (check slice simulation)
- Bridging settings (Speed, extrusion, angle)
- Need more filament cooling
- Material qualities

FIGURE 10.7 – This is a bird's nest. Jonathan Torta.

## › 10.3 – Where can I go for troubleshooting help?

The 3D printing community is vast, with lot of online support and tutorials. If you have a question or need to troubleshoot, chances are there are guides or online help you can learn from.

Here is a list of online guides (at the time of printing) for troubleshooting.

**3D printing for beginners:**

https://3dprintingforbeginners.com/troubleshoot-3d-printing-problems/

**Simplify3D's Print Quality Troubleshooting Guide:**

https://www.simplify3d.com/support/printquality-troubleshooting/

**Simplify3D's Materials Guide:**

https://www.simplify3d.com/support/materials-guide/

**Rigid.ink's poster Advanced Overview to Improve Your 3D Print Finish Quality** (shown at the right):

https://rigid.ink/blogs/news/advanced-finish-quality

**Rigid.ink's Ultimate 3D Print Quality Troubleshooting Guide 2018:**

https://rigid.ink/pages/ultimate-troubleshooting-guide

**Ninjatek's print quality and troubleshooting guide:**

https://ninjatek.com/resources/print-quality-troubleshooting-guide/

**MatterHackers' 3D Printer Troubleshooting Guide:**

https://www.matterhackers.com/articles/3d-printer-troubleshootingguide

**3DVerkstan's Visual Ultimaker Troubleshooting Guide:**

https://support.3dverkstan.se/article/23-a-visual-ultimakertroubleshooting-guide

**All3DP's 3D Printing Troubleshooting guide:**

https://all3dp.com/1/common-3d-printingproblems-troubleshooting-3d-printer-issues/

**RepRap.org's Print Troubleshooting Pictorial Guide:**

http://reprap.org/wiki/Print_Troubleshooting_Pictorial_Guide

**3DVerkstan's Visual Ultimaker Troubleshooting Guide:**

https://support.3dverkstan.se/article/23-a-visual-ultimakertroubleshooting-guide

## SUMMARY

Always calibrate your printer so you don't get poor or unexpected results, and record the settings, any calibration print, and changes you make to help you understand your printer. The Mow Cal test calibration model used in this book includes boxes, spheres and cones, raised and embedded test, holes curves, arches, overhangs, hollow objects, various angles, and detail. Both the level of the bed and the nozzle height are very important. The issues when printing can have a variety of solutions. Only work on one at a time because most settings affect other and modifying more than one can give unpredictable results.

## APPLYING WHAT YOU'VE LEARNED

1. Continue making your own 3D dictionary by adding the definition (in your own words) using five words related to 3D printing in this chapter.
2. Explain the importance of calibrating your printer correctly.
3. Summarize the Mow CAL test in the book and why it is important to test your printer.
4. What happens to your print when the nozzle is too high or low, and what can you do to fix the problem?
5. If you made a 3D object, answer all the questions under "Things to look for" in this chapter.
6. Take 15 of the problems that can occur with 3D printing and give examples of troubleshooting the solutions.

# CHAPTER 11

# Buying and Calibrating Your Filament

## OVERVIEW AND LEARNING OBJECTIVES

**In this chapter:**

- 11.1 – A quick guide to picking filament
- 11.2 – Now that you have filament, what is next?
- 11.3 – Why is printing a benchmark slug important?

In this chapter, we'll explore buying the appropriate filament for your printer and calibrating it by a couple methods. We'll also guide you through printing benchmark slugs and maker coins, which are invaluable objects for recording printer, temperature, and filament settings for your records.

# › 11.1 – A quick guide to picking filament

The filament is the material you will be using for all your 3D prints on an FDM printer. This makes it a very important component for your project. As shown in Chapter 6, there are many types of filament, ranging from basic to hybrids. In addition to picking the type of filament that best fits your project, selecting the size and quality is a big decision that directly affects your print.

## Pick the correct filament for your printer

### Size

Currently the two main filament sizes are 1.75 mm and 3.0mm. DIY enthusiasts started using 3.0mm filament; the thinner 1.75mm and it is now becoming the most popular. Select the filament size that matches the requirements for your printer.

### Compatibility and Limitations

Your printer may have some limitations on what filament it can utilize. For example, most printers can handle PLA ,as it requires the least number of features and generally is the most forgiving.

However, some filaments need additions or lend themselves to certain printer builds. For example, the addition of a headed bed expands the filament selection greatly. Filaments stick to some print bed materials better than others. Bowden extruders and some direct drive extruders with a gap after the drive gear do not handle flexible filament well. Some filament needs to have a more direct cooling system. Some require an enclosure that keeps the temperature from changing too quickly. Some require hardened nozzles due to extreme wear. Some printers use a PFTE tube in the hot end; this degrades at hotter temperatures, thus limiting the filament selection.

It is best to look at your printers' capabilities and filaments requirements. Also, note that because a printer can use a filament, it doesn't always mean that it can handle it in an optimal way.

For example, you don't want to print hotter than 240°C with a PFTE tube. At that temperature and higher it starts to degrade and outgas. You need a heated bed for some materials to stick well to the bed. Some materials require an enclosed temperature-controlled space or strong cooling.

## Pick the correct filament for your project

### Type

Now that FDM printing is popular, there is a very large selection of filament currently on the market, with more being added every day. See Chapter 6 for a list of types of filament.

First and foremost, you need to pick the type of filament. This is a combination of your printer's capabilities (many printers can only print PLA, lower temperature filament, or its extruder can handle flexible filament or a nozzle that may not survive printing with a particularly abrasive filament).

Color can be a consideration if you are not going to paint or cover the print. There is a wide range of colors possible and even factors like glossiness can be selected. Note that not all filament types have this range.

Next is the project's physical qualities: what degrees of strength, flexibility, and durability do you need? If the final use is an art object, these factors may not be paramount. If it will be part of a device or load-bearing, these qualities are very important.

As stated, PLA has become the go-to for the average print. Different formulations provide a wide range of colors, strengths, and even interesting fillings. PLA also tends to have the lowest average price. All in all, PLA is a strong performer and most common. Your project may need qualities that PLA does not have: temperature resistance, flexibility, strength may all require different filament (see **Figure 11.1**).

### Quality

After you select the filament that best fits your project and works with your printer, you need to purchase it.

It can be hard to find a filament that suits your needs and is reasonably priced. Like any other product, some brands are known to be "good" but also have multiple lines of products, including a low end that may not have all the acclaimed qualities of the other lines. Other brands may have good and bad batches – it's the luck of the draw. Still others are consistently bad.

FIGURE 11.1 – 3D printing materials. Maurizio Pesce.

! TIP

A great rule of thumb is to not just shop for the cheapest filament! It could cost you more in the long run.

Some important things you need to be cognizant of when choosing filament:

- Diameter and consistency
- Formulation, impurities, viscosity, and debris
- Moisture content
- Too good to be true, and potential age issues

## Filament diameter and consistency

If you have a printer using 1.75 mm filament, you want filament whose diameter most importantly stays the same through the entire roll. You should be looking for dimensional accuracy between ±0.02 mm and ±0.05 mm (the lower the better); beyond ±0.07 mm can get problematic.

The main problem is that when the filament is radically too large, it will simply jam; too small, it may no longer feed. More commonly, a slight change along its length changes the amount of plastic flowing out of the hot end. This causes uncontrolled under- or over-extrusion, affecting the quality of the print.

### The formulation of the filament thermoplastic is also very important

Currently there is a lot of experimentation in formulation, mostly for the good (strength, temperature range, look), but some manufacturers are also cutting costs. Impurities, some fillers, and even debris can lead to poor molten plastic viscosity, at best compromising your print quality, but also often plugging up the extruder nozzle, resulting in a long cleaning process and a ruined print.

### Moisture is a concern for most thermoplastics

You want good packaging, as the plastic absorbs moisture from the atmosphere relative to the humidity and duration of exposure. When you buy filament that is not in a vacuum-sealed package, you have no idea what condition it is in. Expect the worst and do not use it! You may be able to condition it.

Filament with too much moisture will be evident: it spits and even splatters out of the hot end as trapped moisture is turned into steam as the plastic melts. This is bad for layer adhesion and print quality. It also causes bubbling, which randomly changes the extrusion rate. This results in a printed object that lacks strength and quality. In extreme cases, it could potentially damage your hot end.

One solution for storing filament after you have opened the package is to place it in zipped plastic bags or plastic containers with a desiccant (silica gel pack). This is an absolute must with nylon filament. Some makers create sealed containers that they can feed the plastic out of to keep it isolated from environmental moisture.

### Recovering moist filament

You can recover moist filament by baking your filament in an oven at ~40°C for a few hours (try six hours at first). For PLA, what might work better is a dehydrator (again, no hotter then ~40°C). While some filaments can take more heat, the spool might not, and you do not want the filament deforming.

Note that a common cheap food dehydrator is perfect for this task, and modified versions are being used as dedicated filament dryers.

Alternately, place the filament into an airtight container with plenty of silica desiccant. Then leave it sealed in the airtight box for two to three weeks. This is slower, but it is a sure way to not deform or damage the filament.

It is a best practice to keep the filament in a dry location, constantly sealed, with a desiccant.

### The "too good to be true" and age issues are often linked

A well-packaged filament container should last almost indefinitely. Any damage to that package and time of exposure only degrades it. Additionally, temperature can play a role in deforming it. The older the filament is, the more potential for compromised storage and degradation. So, if possible, use quickly turned-over filament, as it is most likely fresh.

Other issues include entire production lots of plastic that are contaminated, have incorrect formulas, are inconsistently extruded, and so on. These defective products may be sold and dumped onto the market (usually off-brand) at a very low price.

Try filament samples. Often filament producers offer sample packs for a low price. Find what works for you: find a filament that is consistent and stick with that as your baseline go-to filament. Try new filaments – but be picky about quality.

Lastly, use filament that is appropriate. Do not use a costly professional PLA formulated for strength for a non-structural print, when a normal PLA would work for a fraction of the price. Use cheaper filaments for test runs and proto-types. But do use the better high-quality filament products for finished and final prints, as they can make a real difference.

## › 11.2 – Now that you have filament, what is next?

### Calibrate your filament

After you have picked and purchased your filament, you need to set your slicer to the correct settings.

To start, use your printer's best/default calibrated settings for that material (bed temperature, fan speed) and then add to or override the manufacturer settings. This is most often the temperature for hot end and flow rate.

In most cases this will work just fine. If you want to go the next stage and really dial it in, you can. Luckily, calibration here is relatively simple.

### Calibrate for Temperature

The manufacturer will recommend a temperature range. Normally, if you are consistently within this range, you are fine. To combat stringing and filament temperature issues, I usually lean to the lower end of the range. But each printer is different. Use the quality guide in Chapter 10 for issues.

You can print a basic test block or Mow Cal Cat and inspect it. Most of the time it will be in the ball park and only require the smallest changes. Note that 5-degree changes or so are just fine.

Alternately, you can print a temperature tower (see **Figure 11.2**).

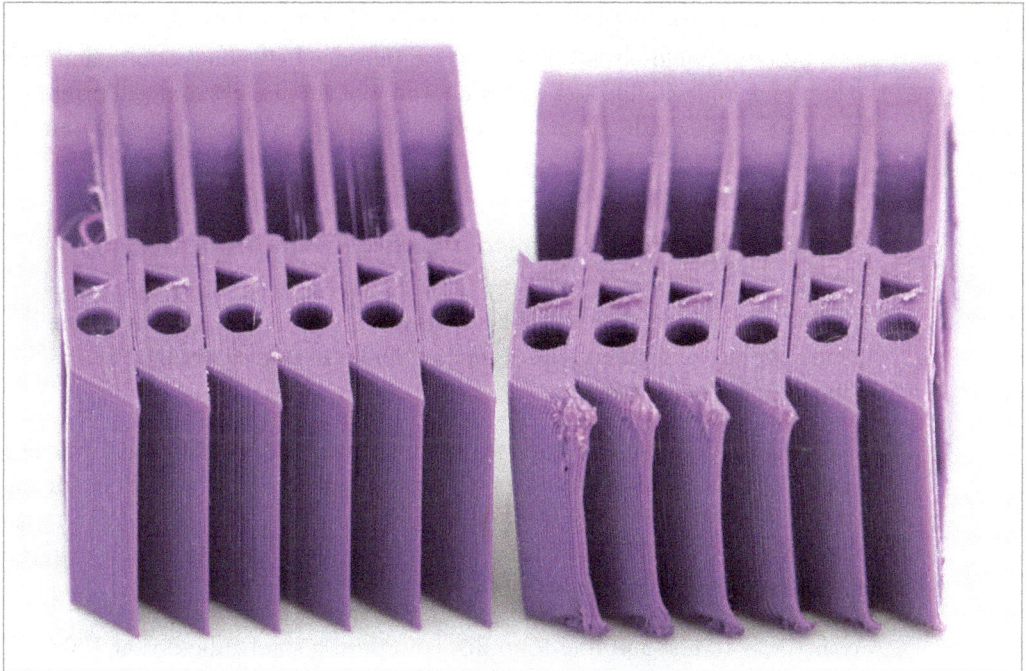

FIGURE 11.2 – An example of a temperature tower. Jonathan Torta.

A temperature tower allows you to gauge the effects of temperature over a range of settings. The example in **Figure 11.2** has a temperature range of 30 degrees overall, in 5-degree increments.

A word of caution: If the initial temperature is too low, you may not be able to even start printing. Make sure to start inside your recommended range! Test this by priming your extruder (extruding a small amount) to ensure the material is flowing.

To start at any temperature on this tower you want, just sink the tower lower into the build floor of the slicer up to the number you want. Anything below that level does not get printed (in most slicers), a quick and easy way to print part of an object.

**MAKER'S NOTE**

The next step relies on your slicer. Two slicers that make this next step relatively easy are simplify3d and Cura, as they both can change print settings, like temperature, versus the number of layers or height in millimeters that are printed. At each step, set a new temperature entry, slowly increasing it as needed. Verify the capabilities of your slicer to see if you can utilize this ability. Then slice your tower and print it out.

Inspect the tower closely, see if there is a difference between sections, and pick the best one. You will find some filaments are not very finicky and will print fine over a wide range (with adequate cooling) and others may need to be within 10 degrees.

### Flow rate calibration (also called the extrusion multiplier for some slicers)

The flow rate or extrusion multiplier setting can be important to fine tune the overall flow rate of filament out of the nozzle; for example, a range from .8–1.0 or 80%–100% of max flow.

The flow can be fine-tuned and will be different between plastics; it can also change between filament brands and formulations. I have found that, for my setup, PLA most often needs a number close to 0.9 and ABS closer to 1.0, with other filaments in between. I have had various formulations of each material that differed greatly, so always test.

A coarse way to test and fine-tune this is to print a solid cube/box at 100% fill and check if there is too much or little plastic.

- Under-extrusion occurs when the printer does not extrude enough plastic and you see gaps between perimeters and infill. Increase the flow rate multiplier.
- Over-extrusion occurs when your printer extrudes too much plastic, resulting in prints that look very messy, gloopy, and overstuffed. Decrease the flow rate multiplier.

## Calibrate by Using Calipers

This method of filament calibration requires a set of calipers. Calipers are an inexpensive tool and a good addition to your toolkit. (See **Figure 11.3**.)

Use this object ***Calibration box*** on DVD with slicer settings that include at least two perimeters/walls, 0% infill, and no top solid layers (set to 0). This is important, as this box would normally print solid. We are tricking the slicer to print the

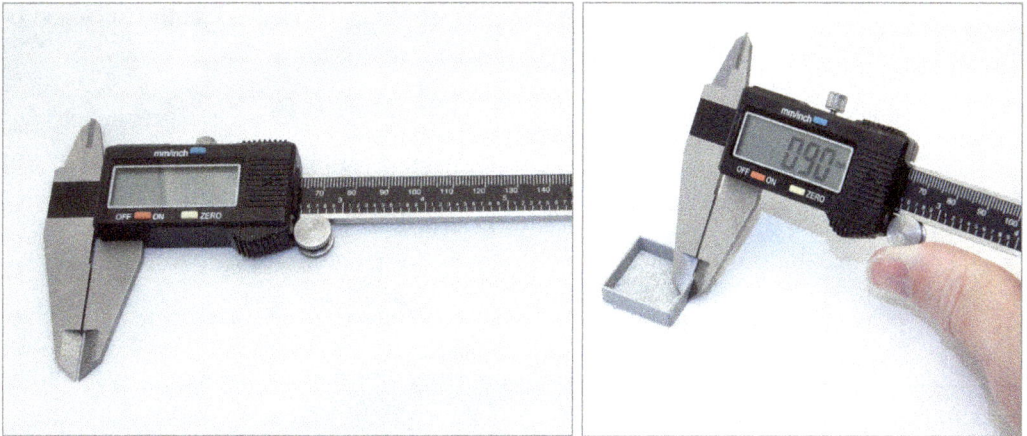

FIGURE 11.3 – Basic calipers. Jonathan Torta.

box with no top or infill to make measuring the wall thickness easy. (See **Figure 11.4.**)

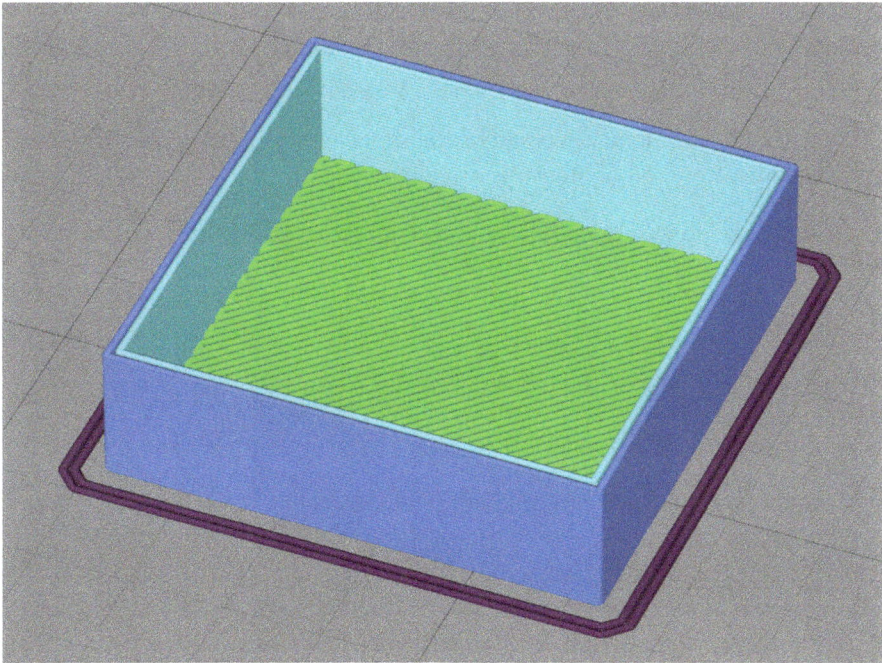

FIGURE 11.4 – Here is a model. On DVD. Jonathan Torta.

Once the object is printed, measure the wall thickness and see how it compares to the extrusion width setting in your slicer.

Ultimately, the goal is to have a single extruded wall/shell ~1.2x the thickness of your nozzle diameter, or 120%.

**A handy equation is:**

New extrusion multiplier/flow = (Expected wall thickness ÷ Observed wall thickness) × Current flow rate

Thus, the extrusion width measurement for a 0.4 mm nozzle would ideally be .48 mm for one shell/wall, .96 mm for two, and 1.44 mm for three.

MAKER'S
NOTE

It is important to measure more than one shell at a time, as this will also incorporate the shell overlap. Given this consideration, I always test two or three shells. Also, if you are a just few hundredths of a millimeter off, you are probably good.

**The variables are:**

Optimal expected wall thickness = (nozzle diameter (mm) × 1.20) * number of perimeters/walls you measure. (As pointed out above, for a 0.4 mm nozzle a multiple of .48 mm)

Observed wall thickness = average of wall thicknesses (mm) of the test object (physical measurement of object using calipers)

Let's set Expected wall thickness = .96 mm for two walls for a 0.4 mm nozzle

Observed wall thickness = 1.04 (measured with calipers from cube print with 2 walls)

Current flow rate = 0.90 as it is the default for PLA in the slicer.

0.87 = (0.96 mm ÷ 0.99 mm) x 0.90

0.87 would be your new extrusion multiplier/flow!

Try another print with this extrusion width setting and see how it behaves. If it prints well, take a measurement. With the flow rate, err on the side of caution – slightly more (a higher number) rather than slightly less. Under-extruding can cause many structural issues, and slight over-extruding can be hidden inside a non-100% infilled print. Also, extrusion will vary slightly – again, better to have it slightly higher then slightly lower. So, don't chance very small changes and try to over-calibrate here.

While the ideal is a hard number, the reality is the filament is more often slightly too small than too large. Too large means a jam. Good manufacturers know this and change their extruding dies often to keep them the correct size, as changing the flow rate is much better than clearing a jam.

Additionally, even when you over-extrude a small amount, most models have infill that is less than 100%, as in open interiors that are not solid. This allows some small flexibility in this direction. Conversely, under-extrusion has much more dire consequences to gaps and layer adhesion. Therefore, I always round up or err to the over-extrusion side a small amount.

MAKER'S NOTE

If the new number works, your printer is calibrated. If not, repeat.

For some fun, print a top for your extrusion tests so you can make them into handy boxes you can use.

https://www.thingiverse.com/thing:3007263 (In extras and on DVD)

MAKER'S NOTE

Another setting that can be changed is the filament diameter. This should be set to your printer's filament size, and if you have a printer profile specifically for your printer, it should be set to 3 mm or 1.75 mm.

However, this can also be set per filament and can be set manually by measuring the filament's true diameter with calipers and inputting that measurement. Be careful, as that can now throw off your flow rate, and small changes here can have a larger effect than just the flow rate.

I find that changing a single variable is much easier to troubleshoot. So, I set the filament's diameter to the reported size and then use the extrusion multiplier/flow to adjust everything; it has much finer control.

MAKER'S NOTE

Now that your printer and filament are calibrated, it is important to create a test object or slug and write down these settings!

## › 11.3 – Why is printing a benchmark slug important?

Creating a benchmark slug provides the benchmark for your printer with the current filament. With this baseline, any change in material or modification to the printer will be more identifiable and trackable.

In FDM printing, there are quite a lot of variables to track. Again, with most slicers, people use the generic profile for your printer and for the filament. This is just fine for basic printing and starting out.

But often a generic profile is just that – generic – and doesn't cut it. Individual qualities of your 3D printer, coupled with the environment and the filament that you're using (the brand/batch, formula, etc.) can change the required settings for a good or even a better print. This also allows you to change settings as needed, uncover advanced settings, and do prints that basic settings can handle.

For example, when selecting a new material to print for the first time (like a new brand of filament, a new blend, or a completely different plastic), I have to determine not only the baseline settings to allow printing, but the ideal settings for my individual printer.

Settings such as flow and temperature are important basic differences that can change per filament. Keep track of advanced settings like retraction distance, print speed, extra start distance, retraction vertical lift, retraction speed, coasting distance, wipe distance, fan start, etc.

I cannot stress this enough – keep a log of every print you make with at least the basic settings and, ideally, what you changed beyond the default settings. This is one of the most important things you can do as a DIY 3D enthusiast! Even experts keep a log of the very minimum setting for different filaments and printers. If you do not have an environmentally controlled space for your printer, keep a log of the environmental settings of each print.

All of this allows you to understand and quantify how changes in basic or advanced settings can affect other settings, how environmental changes can affect your prints, and how different formulas affect how your printer works.

This is where benchmark slugs come in. You can print these and imprint or write information onto a representation of a calibrated object.

## A Basic Slug

This slug is just enough to show the quality and includes space to write down pertinent information. This particular slug comes in a few shapes and includes a hole for attaching it to the spool or a string (see **Figure 11.5**).

FIGURE 11.5 – A basic slug. Jonathan Torta.

We have included a file for this basic benchmark slug, which uses a small amount of filament, gives an idea of print appearance, and has space for writing. This is a great basic example and is simple to use.

At the very least, use something like **Figure 11.6**.

FIGURE 11.6 – A basic benchmark slug model. Jonathan Torta.

You can also make something much more interesting, if you are willing to spend a little more time and filament: a Maker Coin.

**IN EXTRAS** *Project files along with images and videos are included in the extras and on the DVD.*

## Maker Coin

This alternate benchmark slug is a Maker Coin (see **Figure 11.7**).

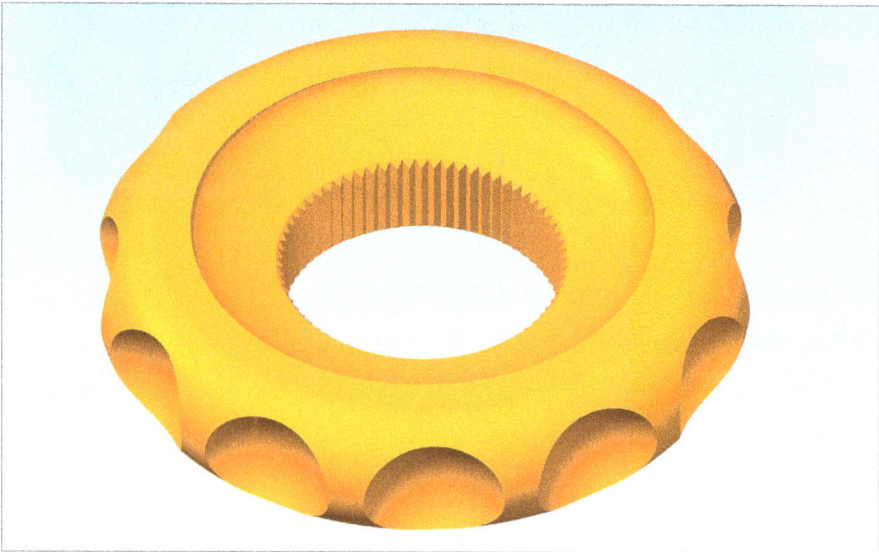

FIGURE 11.7 – A custom Maker Coin. Jonathan Torta.

Maker Coins are like business cards for 3D design and 3D printing. They are often personalized with logos and are often chunky (the size of a large coin). Printing your own Maker Coin is a great first project and an excellent way to test your filament samples. Additionally, if you get into making your own objects, designing your own Maker Coin is great fun.

As you can see in **Figure 11.8**, I made one with the settings I used embossed into the print. Also, on Thingiverse, you can customize your own. I like this because I can print the coin with stats and a reference to my notes.

**MAKER'S NOTE** Write it all down. It is a best practice to keep a build journal and have it handy for review.

FIGURE 11.8 – Example of a Maker Coin, with filament and temperature information embossed on it. Jonathan Torta.

When you find settings you like for a material, it is time to print out a slug. Add that information into your journal and print a slug to keep with that batch of filament. Write or print the relevant information you need and a reference to the entry in the journal. This way, when you use, buy new, or unpack the same filament, you have a defined baseline.

https://www.thingiverse.com/tag:Maker_Coin

Search www.thingiverse.com for "Customizable Maker Coin," open one of the projects, and click on "open in customizer" if available (note you may need to have a free account to access this function)

Web link

https://www.youtube.com/watch?v=tGtNLpYSXOU

## SUMMARY

Filament is the material you use in your FDM printer. Some printers have limitations on what filament you can utilize. There are many reasons it will cost you more in the long run if you use the cheapest filament and don't check its diameter, consistency, moisture level, and age. Make sure you start printing in your recommended temperature range. A temperature tower will help you gauge their effects. The flow rate setting is important to set the rate out of the nozzle. The filament diameter is another important setting. A slug or test object and their settings will help you set a baseline. A maker coin is a great first project and a way of testing sample filaments.

## APPLYING WHAT YOU'VE LEARNED

1. Continue making your own 3D dictionary by adding the definition (in your own words) of five words related to 3D printing in this chapter.
2. Explain the compatibility and the limitations of finding the right filament to use in a 3D project.
3. What are six things you need to consider when choosing a filament? Why are they important?
4. What is a temperature tower and how do you use it?
5. Why is it important to keep a log of calibrations you use on your 3D printer?
6. Explain why a benchmark slug is important.
7. What is a Maker Coin and why print one?

By Frank Schwichtenberg

# Challenges

## OVERVIEW AND LEARNING OBJECTIVES

**In this chapter:**

- 12.1 – My printer is calibrated and I made the benchmark slug – what's next?
- 12.2 – What are some potential issues?
- 12.3 – The challenges!

## › 12.1 – My printer is calibrated and I made the benchmark slug – what's next?

Now that you have calibrated your 3D printer and printed a benchmark slug, you should have an idea of how your printer handles different materials, temperatures, and slicer settings. In this chapter, we will showcase three different object challenges:

- 3D scanned shell
- Sphericon
- Puzzle box

These objects highlight issues encountered during printing and the problem-solving that overcame those issues to create a good print.

**IN EXTRAS** *Project files, color images, and videos are included in the extras and on the DVD.*

Each object has its own unique challenges. Each challenge builds upon the skills you have already learned, combined with additional skills. Let's take a closer look at the challenges.

### The Shell

This object requires support and careful consideration of orientation, layer height, along with sanding and painting (see **Figure 12.1**).

### The Sphericon

The sphericon is separated into two halves, making it much easier to orientate. It requires consideration of the layer height, bottom layer settings, support, raft considerations, cleaning, sanding, gluing, and painting (see **Figure 12.2**).

### The Puzzle Box

This challenge includes a bunch of parts that require correct positioning on the build plate, ideally no support, taken with the bottom layers, and careful cleaning and finishing. And it's fun to assemble! (see **Figure 12.3**).

After working on each challenge, you will gain an understanding of how your printer behaves when printing very different objects and what actions to take to ensure the best outcome for your print.

FIGURE 12.1 – A 3D printed shell and the original. Jonathan Torta.

FIGURE 12.2 – A finished sphericon. Jonathan Torta.

FIGURE 12.3 – A completed puzzle box. Jonathan Torta.

## › 12.2 – What are some potential issues?

Assuming your printer is calibrated, and you have your filament dialed in, you are not done yet! Problem-solving before, during, and after printing are important aspects of the printing process. Additionally, you gain a critical understanding of your print and the filament.

Let's review some important considerations:

1. Orientation
2. Skirts/brim
3. Rafts
4. Infill
5. Layer height

## Let's think about orientation

The orientation of an object on the print bed is very important. Since each layer is printed on the previous layer, you cannot print when an overhang is too great (between 45 and 70 degrees). Any greater, and you will be printing in the air. The filament will not stick to the previous layer and will just hang from the nozzle until it hits a place where it can stick – or nothing – making a blob and strings on your print, or worse, a "birds' nest." This is considered a print failure.

A way around this failure is using a support, a disposable object that the slicer adds to support significant overhangs. It is printed under any overhang, with a slight gap. This allows the filament to have some support but it also does not stick well to the support layers, allowing you to remove the support later.

For example, if you printed a cube on one of its sides, it does not require support and has good contact to the build plate. It is the preferred orientation for this object (see **Figure 12.4**).

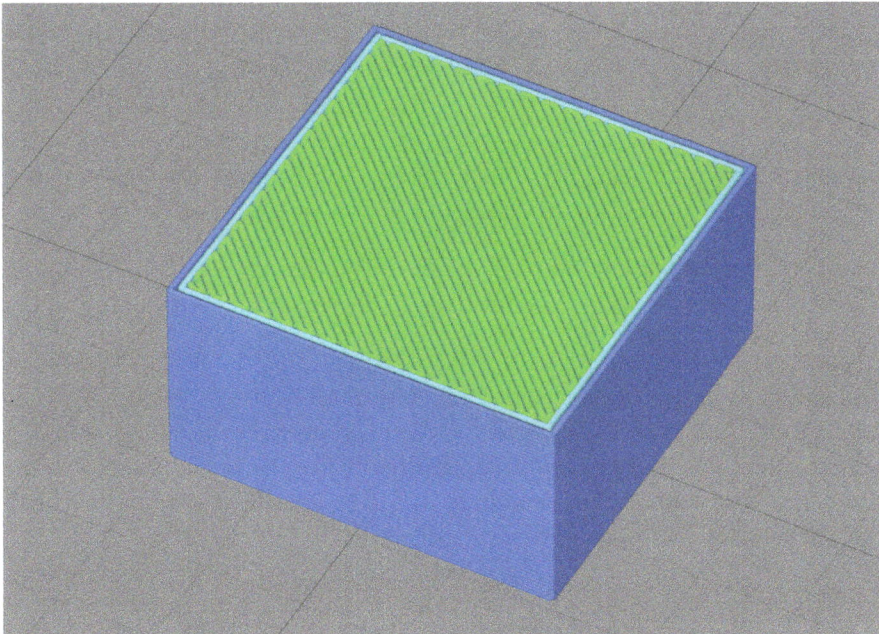

FIGURE 12.4 – A cube, oriented to the build plate on one of its flat sides. Jonathan Torta.

If you tilt the cube up on one edge, it will need supports (see **Figure 12.5**). This orientation is still possible, but it's not ideal, as the sides touching the support will be rougher than the top sides. The top and bottom will also have some stair-

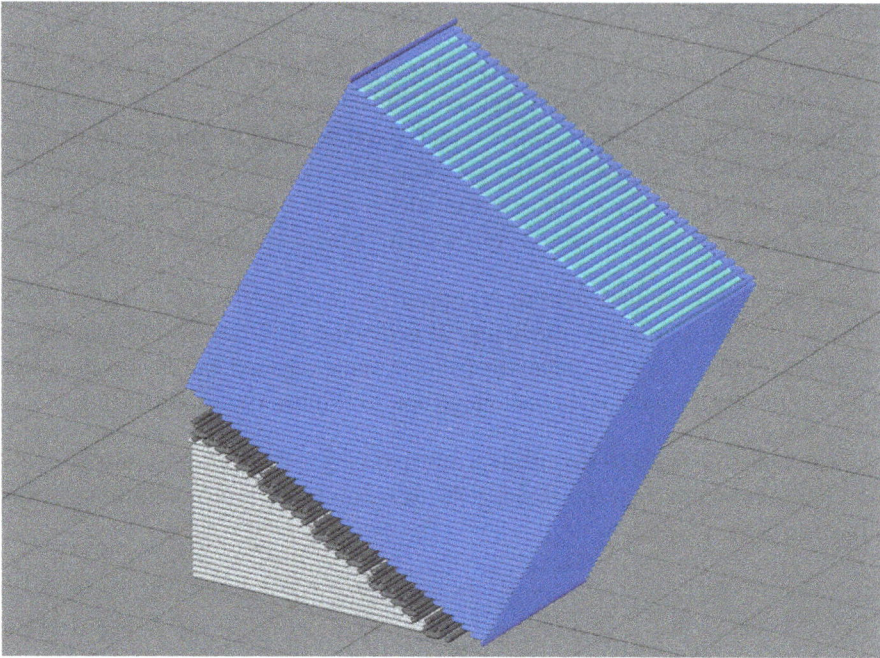

FIGURE 12.5 – A cube (in blue) printed on top of a support structure (gray). Jonathan Torta.

stepping, rather than being flat. Finally, the start of the print may fail because it only touches the build plate with a single line of filament.

Now, let's try printing a sphere. Unlike the cube, there is no side to adhere to the bed and the overhang is greater than 45 degrees for a good portion of the bottom of the sphere.

The answer is to use support and potentially a raft to help with the supports. This not only helps adhere the sphere to the build plate, but also helps form the bottom of the sphere. For more complex objects like the arm of a statue, some carefully placed support may need to be added so the elbow or other low areas, like the hand, are not printed in open space.

Support is very useful and allows you to print objects you would not be able to print otherwise. But normal support is not perfect; the gap that allows for removal also creates blemishes or irregularities on the surface of the print.

Again, orientation is key: put the least-seen sides down, so they use the supports and any surface issues are less noticeable. Use support only when needed.

In Chapter 7 we talked about using the alphabet to visualize what should be supported and what is okay to do without: there is where you experiment.

For instance, "Y" or a "V" has no need for supports but a "T" does. But with some letters it may not be so cut and dried: interestingly, an "H" or an "A" may or may not need supports, since both sides of the horizontal middle are supported, and the printer may be able to bridge this gap, if it is not too wide, with no problem. Wider bridges may sag for a layer or two, but this might also be easy to clean or fix. (So, keep bridging in mind). An "n" may have enough of an arch with that you can print it with minimal sagging also, over even greater spans.

A rule of thumb is, if an object has a part that is horizontal without an end point (overhang) or angles downward, supports are required, as the printer is forced to print in the air. If there is a horizontal span between two areas, this might be a bridge, so check your slicing to see if it will span it correctly, as it needs a foothold on both sides. Arches can mitigate many long spans. Angles can have various outcomes and should be experimented on, as a 45-degree overhang is just fine, while a greater overhang progresses with rougher undersides until it fails completely. When it fails may vary greatly and factors such as material, cooling, number of shells (and order they are printed in), and geometry have an effect.

> If you have a printer that can handle printing with two different filaments, you can use another plastic like Polyvinyl alcohol (PVA) to print the supports. In this manner, there is no need to add the support gap to allow removal because you will dissolve the support in water. This method allows full use of supports that have minimal effect on the object.

**! TIP**

When thinking about orientation, keep in mind that the z-axis is potentially the weakest dimension. Objects can delaminate and break with the grain along the z-axis (very much like wood). A tall, thin object may be more fragile if printed vertically, depending on factors like print temperature, filament type, print thickness, and infill. If strength is a factor, this is something to consider when printing.

Rules of thumb for filament and strength:

- Fused filament is strongest along the axis that it was extruded or deposited. Object dimensions oriented along the x–y axis will be the strongest.
- Bonds between neighboring filament extrusions on the x–y plane are weaker (two extruded lines next to each other). Bonds between layers along the z-axis are weaker still.
- The optimal number of walls or perimeter shells for strength to weight ratio appears to be three (3), as this tends to the best ratio, i.e., most

strength for the least amount of plastic for 0.4 mm nozzles. More walls/perimeter shells tend to add more overall strength to your object than adding more infill.

This seems counterintuitive, but a majority of the stress encountered when bending is in the skin of the object, not the interior. For the same weighted object, one with a thicker skin is always stronger than one with a thinner skin but more internal structure.

- Infill makes an object stronger by adding more material, although its primary use is internal support of upper layers.
- For strength, set the infill angles to be parallel to the axis along which you expect the greatest loads and use a grid or triangle pattern. Most other patterns are not useful for strength. Generally, the higher the infill percentage, the stronger the part. But this is not linear, and at the higher percentages it uses quite a lot of filament for little gain.

Also, parts with a 100% infill percentage do tend to take the most stress before they break, but they also will yield or permanently deform at a lower strain than lower infill percentages. The maximum fill percentage can reduce the quality of the surface of the object. So, an infill percentage setting of 90% or lower may be the best all-around choice when you want a strong part.

- A layer height of 0.2–0.4 mm is somewhat stronger than 0.10–0.15 mm. This difference is noticeable but not as large as previous factors.

## Technical deep dive: Stress-strain analysis and orientation

The basic orientations for loads in stress-strain analysis are: compression, tension, bending (also known as flexing), torsion (also known as twisting), and shear. (See **Figure 12.6**.)

The stresses analyzed are:

- **Maximum stress** is a measure of the maximum stress at the breaking point of the part. It is also known as ultimate tensile strength (UTS), max stress, ultimate stress, or stress at break.
- **Yield stress** is a measure of strength at the point where the part experiences a permanent deformation of its original dimension.
- **Elongation at break** is the percentage that the material is stretched at the point where it fails.

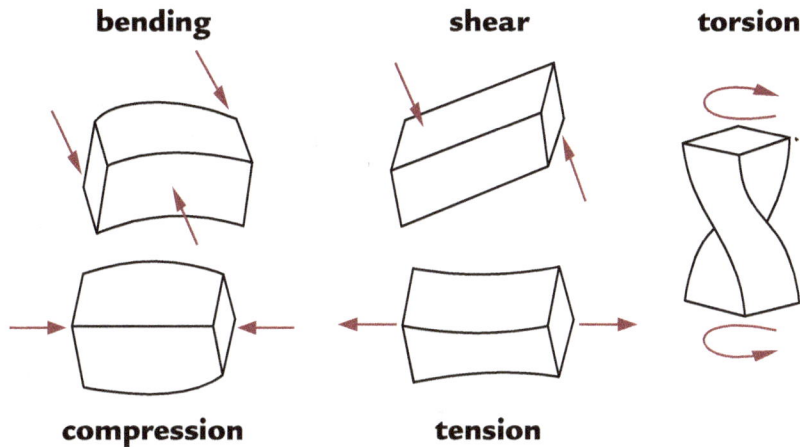

**bending**   **shear**   **torsion**

**compression**   **tension**

FIGURE 12.6 – Stress-strain analysis.

Other stress-strain analysis metrics:

- **Geometric stiffness** is a measure related to the shape of the object. Some shapes have inherent strengths versus a particular load than others.

- **Toughness** is the amount of energy that a material can absorb before it fractures.

- **The Young modulus** (or rigidity) is a measure of the stiffness of a given material and the measure of a material's ability to resist permanent deformation.

- **Harness** is a measure of the resistance that the surface of the object versus penetration by another stronger object.

The general strength of an FDM print is primarily based on the amount of filament used (wall thickness and infill), and as mentioned, this can plateau. As a rule of thumb, the more wall thickness and then infill, the stronger the object. To a lesser extent, the thicker the layers, the stronger they are. Each layer is a thicker continuous filament and fewer layers mean fewer potential issues with delamination. (See **Figure 12.7**.)

Infill can greatly add to the time it takes to print and can add 20%–25% more time when changing from 10% to 90% infill across all layer heights. Time increases for more complex infill patterns upwards to 60+%.

100% infill can affect exterior quality. At this extreme, the filament flow rate calibration must be perfect. Any variation in filament diameter at all will affect the

# Quality/Smoothness

|  | Infill % | | | | | | |
|---|---|---|---|---|---|---|---|
|  | 10 | 30 | 50 | 70 | 80 | 90 | 100 |
| Layer Height (mm) 0.1 | very high | | | | | | |
| 0.15 | high | | | | | | low |
| 0.2 | medium | | | | | | |
| 0.25 | low | | | | | | |
| 0.3 | Very low | | | | | | |

# Strength

|  | Infill % | | | | | | |
|---|---|---|---|---|---|---|---|
|  | 10 | 30 | 50 | 70 | 80 | 90 | 100 |
| Layer Height (mm) 0.1 | | | | | | | |
| 0.15 | | | | | | | |
| 0.2 | | | | | | | |
| 0.25 | | | | | | | |
| 0.3 | | | | | | | |

# Speed

|  | Infill % | | | | | | |
|---|---|---|---|---|---|---|---|
|  | 10 | 30 | 50 | 70 | 80 | 90 | 100 |
| Layer Height (mm) 0.1 | | | | | | | |
| 0.15 | | | | | | | |
| 0.2 | | | | | | | |
| 0.25 | | | | | | | |
| 0.3 | | | | | | | |

# Cost

|  | Infill % | | | | | | |
|---|---|---|---|---|---|---|---|
|  | 10 | 30 | 50 | 70 | 80 | 90 | 100 |
| Layer Height (mm) 0.1 | | | | | | | |
| 0.15 | | | | | | | |
| 0.2 | | | | | | | |
| 0.25 | | | | | | | |
| 0.3 | | | | | | | |

Best ▮▮▮▮▮ Worst

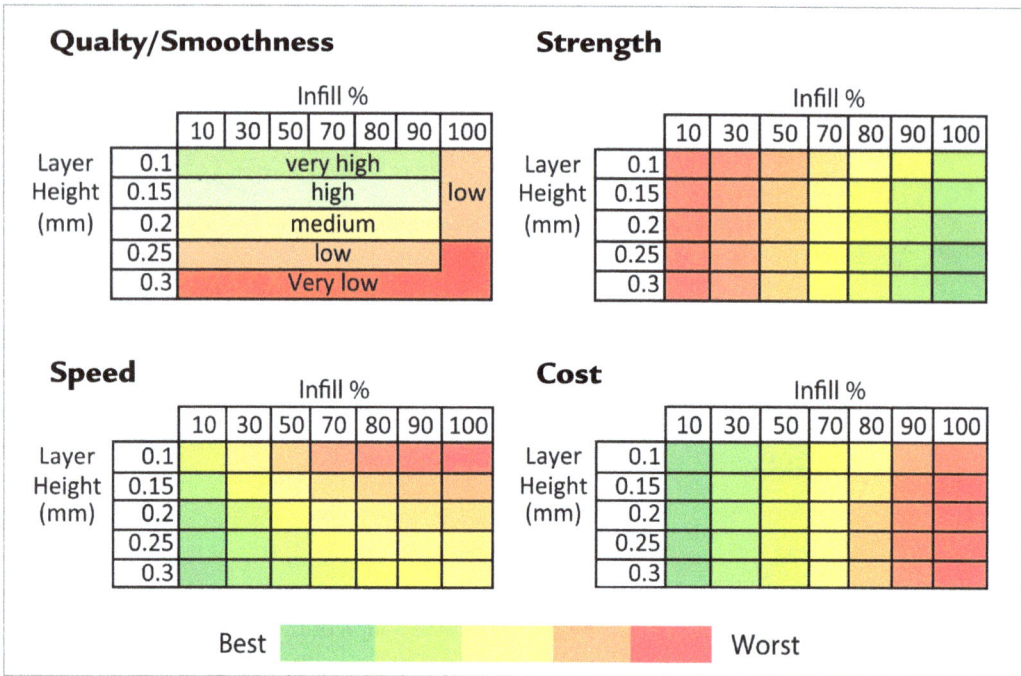

FIGURE 12.7 – Diagrams of the relations of strength, speed, cost, and quality to layer height.

flow rate also. (In Chapter 11). At 100%, there is no space for any over-extrusion to go, as normally there is space between the shells/walls and within infill. This can result in "overstuffed" prints with surface deformations.

Infills of less than 20% can affect top layer quality. As stated above, infill primarily supports the upper layers of an object. Too little infill and any large internal overhang will sag.

For example, take a normal cube: the top is flat without internal support and the top has to bridge the gap. If the cube is small, this is fine, but the larger it becomes, the greater the sag. Infill of 20% normally eliminates this problem. You can use less, but an additional increase of top solid layers is required and may not always work.

Infill patterns are interesting – some slicers have quite a few and it is tempting to use them. For a normal print, stick with linear infill unless you really need/ want something different. Different infill patterns tend not to have much overall good effect and most all non-linear patterns tend to be weaker and slower (from slightly to substantially). There are many factors here; alternate infill patterns can be useful but use sparingly.

Note that some new dynamic infill patterns may be superior in strength and even better in saving filaments but are not in common slicers yet.

Remember the inherent weaknesses along the z-axis, because the interface between layers is not as strong. For infill parts under tension, the z-axis direction can be 20% to 30% weaker than other directions. Again, different materials, temperatures, and even post-processing coatings can modify this factor. It will always be weaker.

Z-axis quality is normally proportional with layer height (the lower, the better). Note that this ends at the limit of the printer's capability. This top-quality limit also includes filament type, nozzle, and even infill. At this extreme, the printer may have issues in reliability; a failed print means wasted time and filament. Also, the lower layer height is slightly less strong. Keep this in mind when selecting settings.

All of this shows that how you place even a simple object not only is important but affects a lot of factors. If strength is a factor, or for mechanical or load-bearing parts, this is very important. For sculptures and most shelf items, this is normally not an issue, unless they have delicate parts.

Orientation, wall/shell thickness, infill, and layer height are extremely important to consider and balance what is needed.

## A practical look at orientation

Now, let's think about all these factors at once for a more complex object.

For example, take a look at this cosplay object. It is a holder for an LED and battery that attaches to the back of a boot to make magboots (magnetic boots) from the television show *The Expanse* by Amazon Video. (See **Figure 12.8**.)

How would you prepare this model for print? What questions should you ask yourself? (See **Figure 12.9**.)

- How do you orient it?
- Do you lay it on its back? If so, would you use much more support?
- Will it matter if the back of the object is rougher?
- How about the details on the top?
- How about the strength of the side loops? Will they hold under pressure?

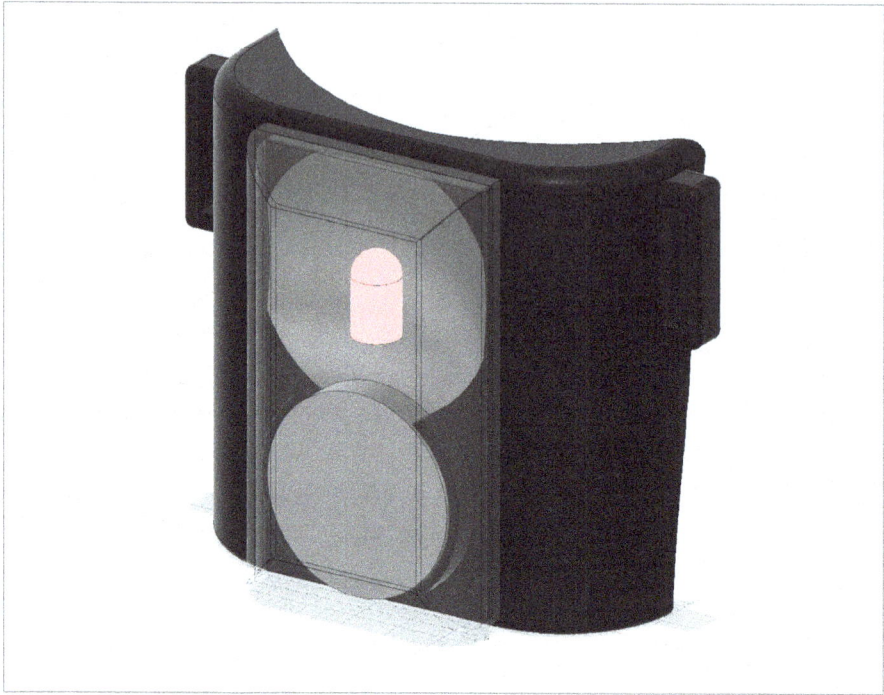

FIGURE 12.8 – A 3D model of the back of the magboot cosplay model mockup in Fusion3d. Jonathan Torta.

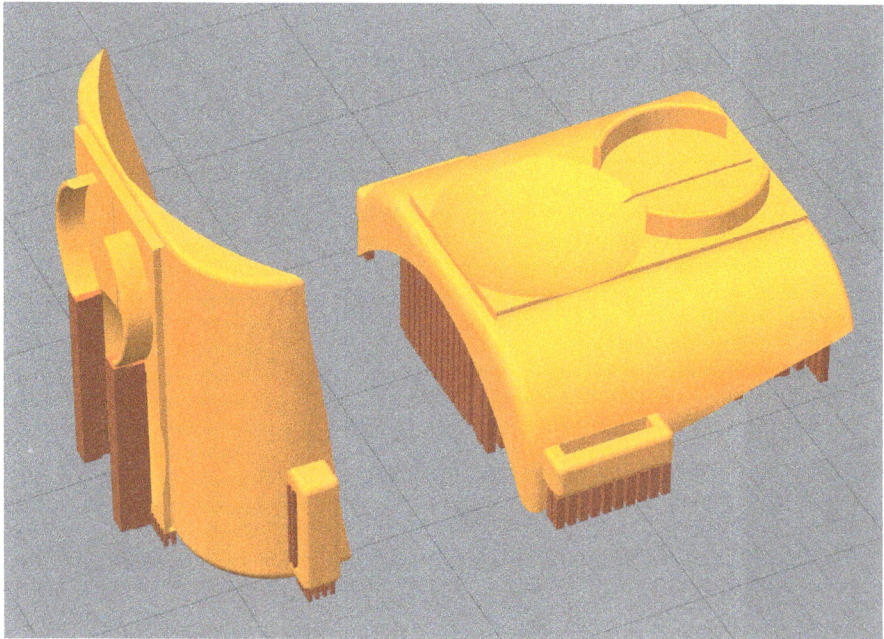

FIGURE 12.9 – Orientation experimentation of the magboot 3D model. Jonathan Torta.

Let's think about this: this prop is a boot attachment.

- The concave back will attach to the boot and will not be seen.
- It will be held there with elastic threaded through two loops on either side of the object.
- The "parenthesis" will hold the button battery and one lead of the LED will fit into the groove.
- A clear lens will fit over the battery and LED.

The important considerations for 3D printing:

- The details are on the lens, battery, and LED side.
- The back will not be seen.
- The side loops will have elastic threaded through them and will hold the unit to a boot.
- It only has to hold its own weight and the shape will keep it in place, so it does not need to be tight.

We need to print this out and we have some considerations to keep in mind.

Here are some points and their reasoning for each side:

## Side (either)

- Some surface area on bed but not a lot (side of loop) – no real benefit.
- A lot of support on one side on a large curved area – this area will be seen. So less than ideal and in this case, it's a deal-breaker.
- Detail on the front should print nicely in the z-axis.
- Stair-stepping along the large side curves and top would be noticeable, requiring more cleanup work.

## Back (toward boot)

- It would use much more support. This is acceptable but there's a potential issue with so much support; may need a raft to ensure the support is correctly printed and secure to bed.
- It would make the side loops the strongest – the best case for loops.
- Top detail (the thin indent) is very narrow – potential detail loss. Need to do a demo slice to check if the detail is not lost, but probably okay.
- Stair-stepping along the large side curves and top would be noticeable, possibly requiring more cleanup work.

### Fat end (top)

- Minimal support, only under side loops and battery holder. Very little surface area to clean up. (Although note that it is the top surfaces of both).

- Side loops are not with the "grain," so are potentially weaker. A potential failure point – check if it is an issue. Note that since these do not require a lot of strength, it should not be a big issue. If so, an alternative is to make the loop thicker.

- In this orientation, most surface area is on print bed; it should make a great base for printing.

- All detail is in the z-axis and should afford the best resolution.

- The only potential stair-stepping is on the thin fillet on the very top and bottom. Should not be noticeable.

### Narrow end (bottom)

- Minimal support, only under side loops and battery holder. Very little surface area to clean up. Much taller support required for side loops (minor, but more chance for tall, narrow supports to fail during print).

- Side loops are not with the "grain," so are potentially weaker. A potential failure point – check if an issue. Note that since these do not require a lot of strength, it should not be a big issue. If so, an alternative is to make the loop thicker.

- Much less surface area on print bed in this orientation than the top but possibly adequate. Might use a skirt if more is needed.

- All detail is in the z-axis and should afford the best resolution.

- The only potential stair-stepping is on the thin fillet on the very top and bottom. Should not be noticeable.

My solution for this object was to print it vertically. I chose to print on the top end down, but the bottom could have also worked. The side orientation was not good solution at all. And the back might have worked but would have had stair stepping along the face of the more pronounced side curves.

A lot to consider, right?

# Let's think about skirts/brims

## Skirt

A skirt aids priming as an initial extra extrusion to ensure the plastic is flowing and up to speed. This should be left on, unless you need a brim or raft (see below). Priming the nozzle is always important and the first layer is critical. An under-extruded first layer will fail. Also, this is an aid to make sure that when you change filament, the printer has extruded enough to fully clear the nozzle of the previous filament (See **Figure 12.10**).

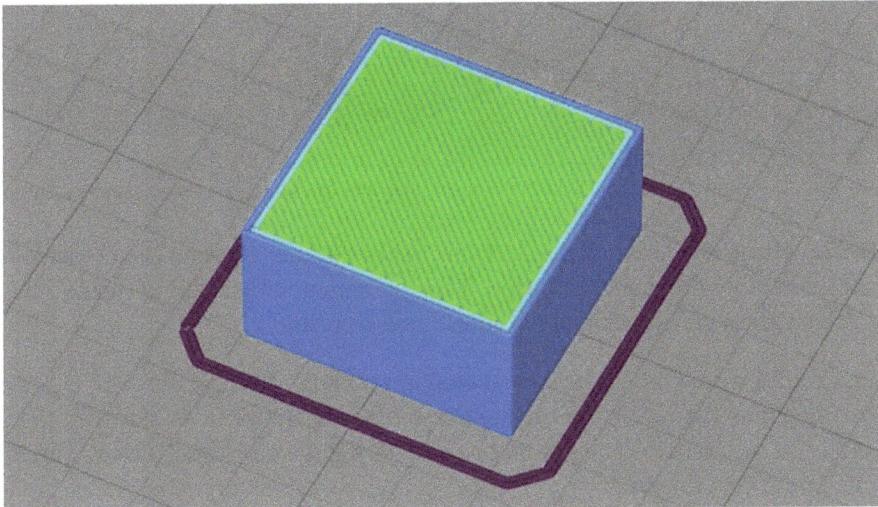

FIGURE 12.10 – Skirt, a few lines around the object to ensure the filament is flowing well. Jonathan Torta.

## Brim

A brim does double duty: it aids in priming and adds additional surface area to help hold an object that may come loose. Brims are quite useful for the extra holding, as they take little plastic and can also shore up areas that can curl due to cooling (for instance, ABS prints). Brims are also quite easy to remove (see **Figure 12.11**).

## When to use a skirt or brim

**Material flow and priming:** At the very start of a print, the plastic flows inconsistently due to incomplete melting or melted filament oozing out of the print head prior to the start. A skirt is very useful to position the print head and normalize flow before printing that very critical first layer.

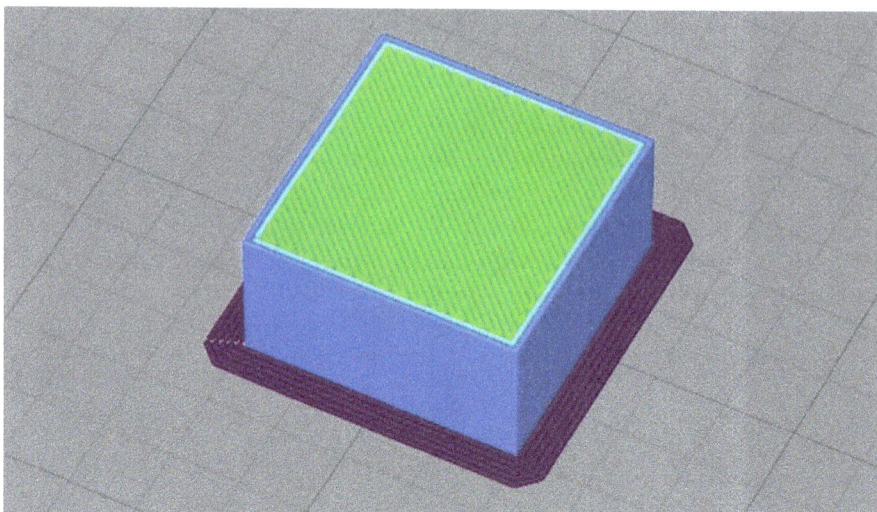

FIGURE 12.11 – Brim, a skirt that contacts the object, helps hold and is easily removed. Jonathan Torta.

**Bed leveling:** While printing a skirt, any issues with bed leveling can be visually detected before the object is started. This can help in proactively correcting the problems through a quick manual adjustment or restart.

**Layer adhesion:** One of the most common problems during a 3D print is layer adhesion. The first layer can dictate the success of the print. A skirt can help hold corners or narrow parts of an object that might come loose. This is a great solution for thin towers.

---

!
TIP

The larger the brim, the easier it is to remove. They normally come off in a single sheet or peel. I often use a pair of needle-nose pliers to help remove it. If you only have a brim a few shells thick, it is much harder to grab and it might rip before being fully removed. Do not skimp on the brim!

Additionally, these are lifesavers for temperamental prints that like curling. Often curling starts and is worse at corners. A brim supplies much-needed surface area on sharp corners to hold them down.

---

## Let's think about rafts

Rafts were originally intended to ensure good adhesion and flatness for the first layer. The first printers had some real issues with adhesion, leveling, and even bowing of the build plate. These issues have not disappeared completely, but they are much less of an issue than they once were. Thus, a raft is often no longer required, but it is still a great tool to use when needed.

The raft is basically a 3D printed base to print an object on. The first layer is printed slowly and thickly for maximum adhesion and clearance from the print bed, then a few more layers are added, each tighter together, to make a good base to start the object. There is a small gap to control the adhesion to the raft (this controls how hard it is to remove) and this *can* make the bottom of the object slightly rough. A raft is also a great, strong base for support and is often use to ensure the support does not pull off from the build plate (see **Figure 12.12**).

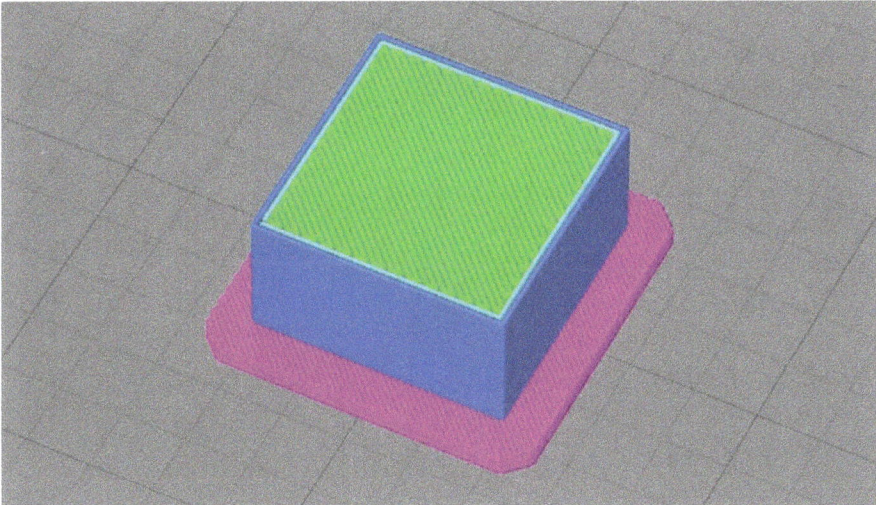

FIGURE 12.12 – A basic raft underneath and extending from under a simple box object. Jonathan Torta.

For some complex or temperamental projects, a raft may be something to consider. They are more robust than a brim. Rafts hold better (due to increased surface area) and have the benefit of not being affected by print bed unevenness or mis-calibration (to a reasonable extent.)

Most current printers can get away with no raft for most prints. This saves plastic and printing on the bed directly can leave the bottom face very smooth.

A raft is most useful under the following conditions:

- An irregular object that has a preferable orientation that does not include enough surface area to hold it on the bed, used in concert with support.

- A large object that fills the entire print bed, as even the best leveling may not be good enough for a pristine first layer.
- Where warping may be an issue.
- When the layer height is very small. A small layer height exaggerates issues in the leveling and other first layer issues. A raft starts with a large layer height, removing those issues.
- You may not want a super smooth bottom layer. You may want a side more like the other sides for post-production or gluing. This also depends on your print bed, as some are textured and some are smooth.

Most rafts have a few parameters to note, but default settings typically are best.

Basic parameters:

- **Top layers** – density and number
- **Bottom layer** – infill, height
- **Offset from object** – size larger than the footprint of the object
- **Separation from object** – air gap, the larger the easier to remove but lighter hold
- **Print speed** – normally slow (~50% normal speed)
- **Similar settings** for middle layers if they exist

## Let's think about infill

Infill, as mentioned earlier, is a way of filling the interior volume of an object. Normally this is also a way to decrease the amount of plastic used by only partially filling this space.

The two main parameters for infill are the pattern used and how dense that pattern is. Each style has its own strengths and weaknesses, and each its own uses (see **Figure 12.13**).

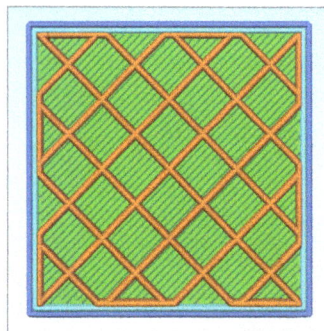

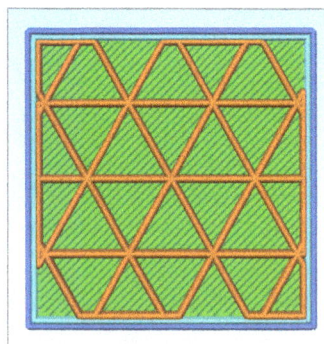

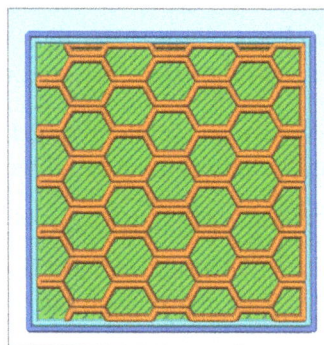

FIGURE 12.13 – Basic infill patterns at different infill percentages. Jonathan Torta.

- **Recliner infill pattern** – good support and effectively the best for strength and speed.
- **Triangular infill pattern** – good support, good speed.
- **Honeycomb infill pattern** – slow but good in support of top layers and strength

---

Different slicers handle infill percentage differently. So, a recliner/grid pattern may look denser than the triangular. You really should try a few test prints. Since infill is primarily for support, the patterns with the least diagonal gap (standard top fill is diagonal) using the least material and printing fastest are normally the winners. If you add strength into this, geometric design also should be taken into account. With new patterns and even dynamic and 3D patterns now coming out, there is quite a lot to test.

---

Since infill takes up the space inside a print, it makes sense that infill designed for structure works better than infill designed for aesthetics. In this case, rectilinear or concentric patterns incorporating grids and honeycombs work best. Grids tend to print fast. Triangular and honeycomb give good support but are slower. Lines, waves, random, or animal patterns are nearly useless.

The density controls the repetitiveness of the pattern and the overall density of the plastic deposited.

Common infill densities are between 20%–25% for the recliner infill pattern (remember, these are ballpark numbers, as different patterns handle the percentage slightly differently). This range offers a nice balance between durability and material consumption.

If structure isn't a concern but cost is, the best infill range is between 10% and 15% to save plastic, and more like 30%–60% if you need more mass and structure. On the low side, you must remember that the top of your object may use this structure for support, so too little may affect the top layer quality quite a bit because the extruded filament needs to bridge the gap and is especially notable if you have few top layers.

On the other end of the spectrum, you can have a totally solid object at 100% infill, but make sure the printer has a perfectly tuned extrusion rate to the plastic you are using, as any minor over-extrusion has nowhere to go and will harm the look of the exterior.

Some slicers have different abilities here, like dynamic infill. This slowly changes the infill from a low percentage to a high percentage near the top of the print, fixing the bridging issue of having too little infill support and still saving material. They may have the ability to use different settings at different heights (for instance, start at 10%, mid-way change to 20%, and near the top switch to 30%).

For some new, interesting patterns to try, see **Figure 12.14**.

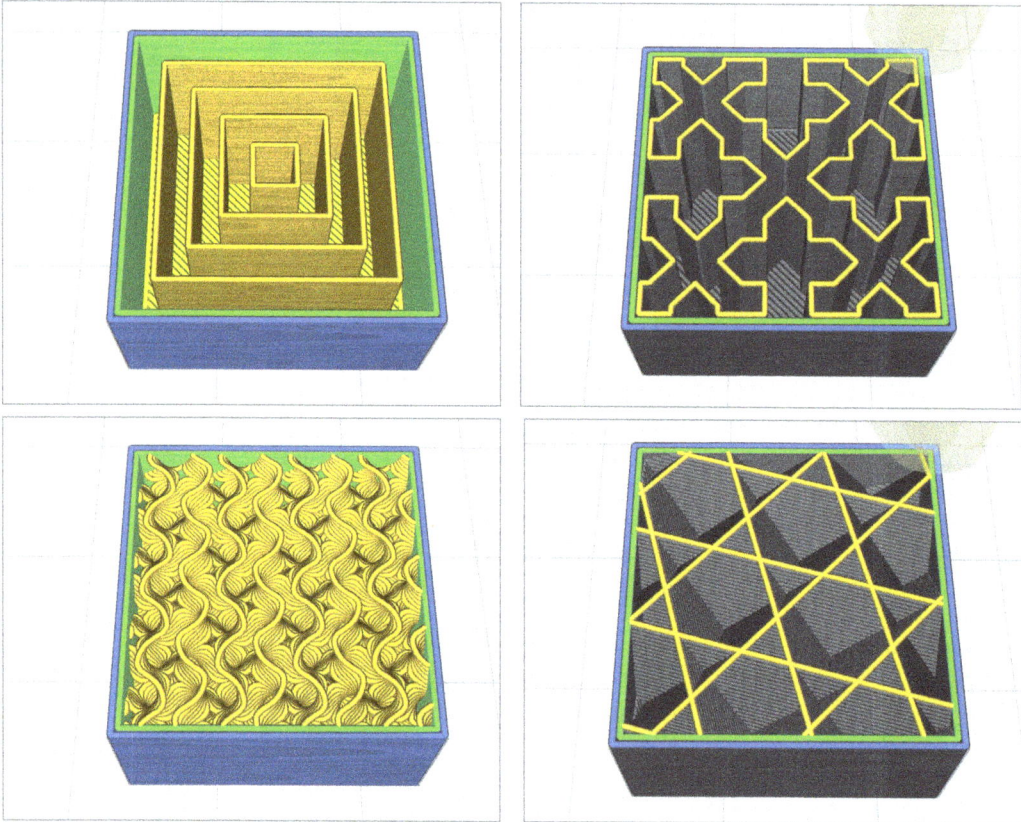

FIGURE 12.14 – New infill patterns: concentric, cross, and two 3D infills: gyroid and cubic. Jonathan Torta.

## Let's think about layer height

As discussed, layer height is exactly as it sounds. More is thicker layers and coarser resolution; less is thinner layers and a finer resolution. As seen previously, this also affects print time greatly; moving from 0.2 mm height to 0.1 mm height doubles the number of layers, forcing the printer to do two times the motions to finish the object.

Additionally, your printer will have a minimum useful layer height (one that can be used to reliably print) and an upper limit of a percentage of nozzle diameter. As mentioned earlier, your print bed may affect the minimum possible without a raft. Not to mention that all plastic does not flow the same way, as some are fine, with a very thin layer, and others require a much larger bead to flow out of the nozzle or potentially jam the extruder.

Another orientation issue, discussed in Chapter 7, is stair-stepping and x/y resolution. So, we know the average FDM nozzle diameter is .4 mm. This is a limit of resolution of some details in the x/y plane (parallel to the build plate.) The vertical axis (the z-axis) is much more controllable and can be typically set from 0.3 mm to ~0.1 mm. Potentially you can have four times the resolution on the z-axis. Also, the smaller the layer height, the less noticeable the stair-stepping topographic map effect you will get at the top of your print.

For example, you may have a tall rectangle with embossed detail on it. It may make the most sense to place it on its back so it has good attachment to the build plate. But the fine detail is on the x/y plane in this orientation and if it is less than ~0.4 mm–0.48 mm (with a 0.4 mm nozzle), details could be lost. It is better to print the tall rectangle on the smaller end so the detail gets printed vertically (see **Figures 12.15–18**).

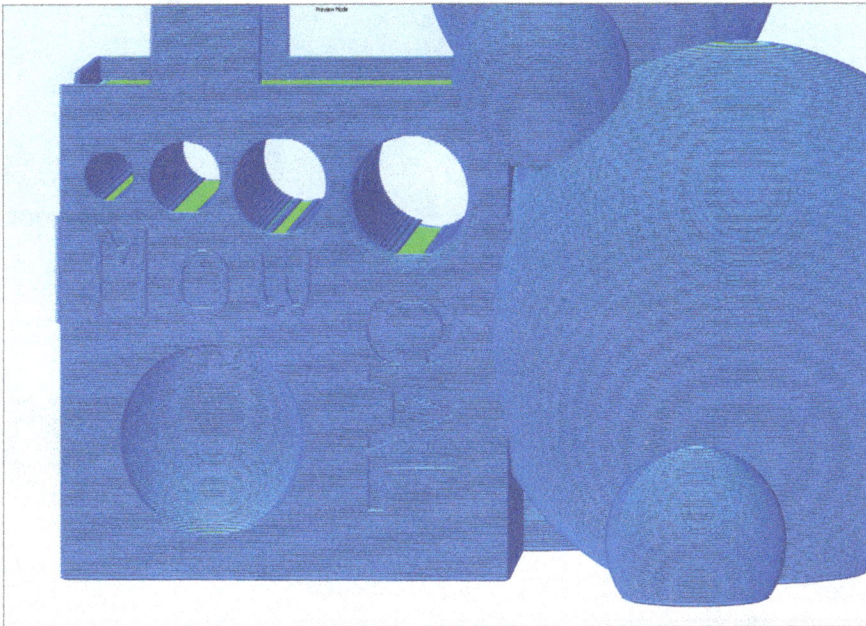

FIGURE 12.15 – Sliced with words vertically (z-axis) with 0.4 mm nozzle. Jonathan Torta.

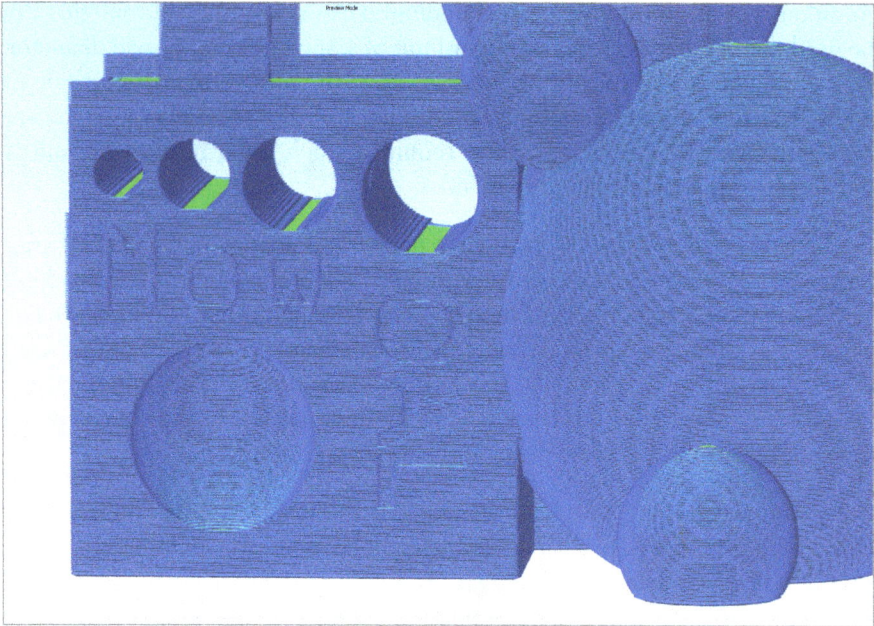

FIGURE 12.16 – Sliced vertically with 0.5 mm nozzle (no loss of raised detail). Jonathan Torta.

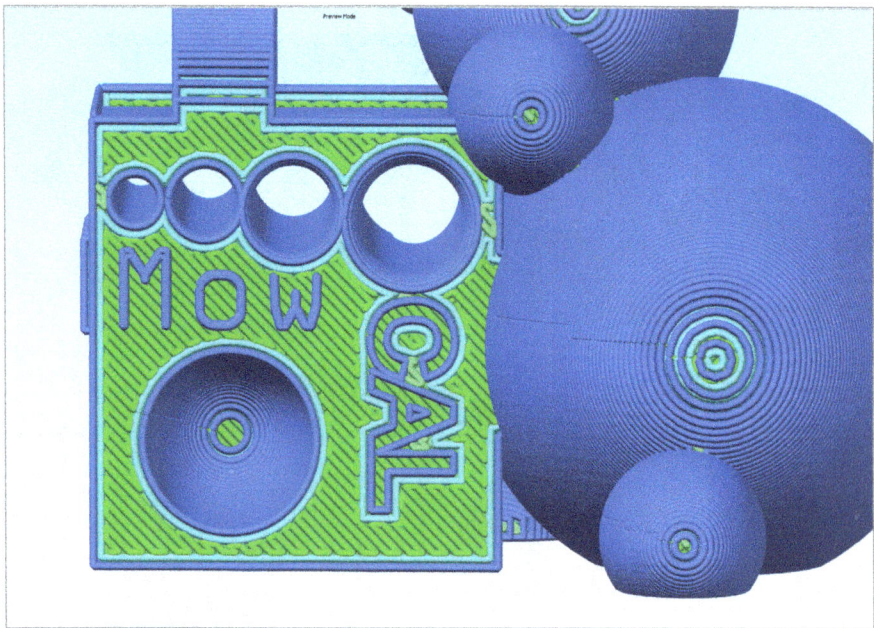

FIGURE 12.17 – Sliced with words "flat" (x/y axis) with a 0.4 mm nozzle. Jonathan Torta.

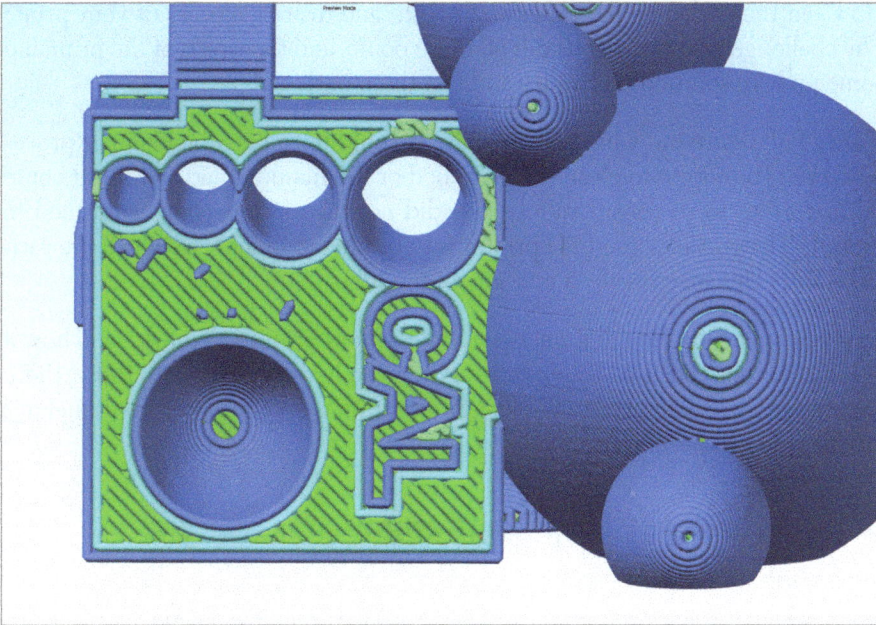

FIGURE 12.18 – Sliced flat with a 0.5 mm nozzle (note the loss of raised detail). Jonathan Torta.

## › 12.3 – The challenges!

Now look at the three objects challenges, on the DVD or website.
- What skills do you want to exercise first?
- How are you going to orient the shell? What makes most sense?
- What sort of settings will be required for the sphericon?
- The puzzle box itself may be a bit of a puzzle to set up:
  - › How do you orient the parts?
  - › Will some parts require support?
- Print all or just some of the parts at a time?

As you may notice, the challenges are more complicated than at first look. There is quite a lot of information to consider.

Remember the important considerations:

1. Orientation
2. Skirts/brim
3. Rafts
4. Infill
5. Layer height

How can these change the quality, strength, and overall success of your print? The challenge is to really consider all these points and the needs of the print and come up with solutions. Then try and test them out!

If they fail, figure out why and try again. It could have been bad luck or procedure (like you forgot to clean the print bed or the filament had a kink). It could be that there were considerations you did not take into account and need to rethink. The goal of a good 3D printer is to be able to quickly weigh all the variables and print successfully a majority of the time.

This takes practice, thought, understanding of your printer, its quirks, and how it works, and these basic considerations. When you become good with these, there are always new challenges to try, like different filaments with different qualities and hard to print objects.

## Suggestions for each challenge

### The shell

Try different orientations. Note how you want the shell to sit (like on a shelf) and what sides will be hidden from view. See about placing the support there.

Try changing the support settings if possible, in your slicer. Try changing the max supported angle and see how that changes the location of the supports, a denser top few layers of support leave less marring of the surface. Does your slicer support manual placing of supports? If so, try that.

This is a perfect object to sand, smooth out, and paint. What layer height will you print it in? How long did it take to print versus sanding and smoothing?

How did the painting go? Did you use primer? How did you paint the shell – spray, brush, or airbrush? Did you clear-coat it?

### The sphericon

This print is much easier with orientations– there is really only one good one for all the parts except one. What layer height did you print this in? Does your printer allow different profiles or automatic layer heights? If so, did you try it?

What did you set your first layer height to? Did you get an elephant foot? If you did, how would you fix it – during print or after print? Did you use a brim or a raft? How well do the parts fit together? Any warping or gaps? Does the key fit?

Are you going to glue the sphericon together? Fill the seam and paint it? Did you sand the sphericon before or after assembling it? Do you think it makes a difference? How did that go? What did you use?

## The puzzle box

The puzzle box has a bunch of parts. Each part is an interesting shape – can you place all of them so you don't need to use any supports? Yes? No? If no, how can you minimize support use? How did you place the parts on the print bed? All at once? Used a built-in command to place? Manually picked what side touches the print bed? Are you going to print these out all at once? In small batches? One by one? What layer height makes sense to use? How long will the entire (and/or each individual) print take? Did you find any place on your print bed that was not even or where the print bed was not leveled/trammed as well as you would like? Are you going to print each or each set of parts in a different color filament?

How are you going to finish the puzzle Box? Are you done with raw but different color blocks? Did you print it all in one color and expect to paint it? Are you going to sand it? If so, what's your plan? It is a lot of flat surface area! How are you going to paint it? Each part a different color or something else, like the exterior faces a uniform color and the "inside faces" another? What other ways can you think of?

Some additional features that some slicers have can be useful to think about and keep in your mental toolbox.

- The ability to change layer height dynamically or incrementally at set heights. This makes printing shapes like the sphericon or any curved shape potentially easier.

  Print vertical sides at a fast and larger layer height but then lower it for the top/curve details. This is a best-of-both-worlds setting, saving time while keeping detail where needed.

- Changing infill dynamically or incrementally at set heights. This too can be useful for prints that do not require infill for strength but to only support the top layers. Start at low infill and then increase it in time for the top layers to be supported by it. This can save time and plastic while also keeping quality.

- Ironing the top layers. This is an interesting feature that moves the hot end over the last printed layer a few times to "heat polish" the last layer and remove the filament lines. (this can leave the top layer looking like the bottom (printed on glass). Nice and smooth. Note this takes extra time.

- Randomly varying the outer wall/shell a small amount. This can give a rough exterior that could be handy for grips, or just to give a softer, fuzzy look.

- Randomly varying the temperature slightly. When used with temperature-sensitive plastics like Wood PLA, this can produce shading of darker and lighter layers making an even better wood look.
- Auto orientation logarithms. These will analyze your objects and find the orientation with the least overhang and/or surface area to adhere to the print bed.
- Functions that would allow you to print different objects with different profiles (settings) all on the same print run.
- Print in a continuous spiral (vase mode) and one continuous shell.
- Stop and start a print allowing filament changes in-between, allowing a multi-color print with a printer that was not made to handle multi-color.

## SUMMARY

The shell, the sphericon, and the puzzle box in this chapter are highlighted to show you issues that can occur during printing. Each of the three projects has its own challenges and skills needed to complete the object. The shell requires support, careful consideration of orientation and layer height, along with sanding and painting. The sphericon requires layer height, bottom layer, support, raft, cleaning, gluing, and painting. The puzzle box requires parts that need correct positioning on the build plate, ideally no support, care taken for the bottom layers, and careful cleaning and finishing. Orientation on the bed is very important because each layer is printed on the next layer. The skirt/brim, rafts, infill, and height also need to be considered. There are many challenges but the finished project is worth it.

## APPLYING WHAT YOU'VE LEARNED

1. Continue making your own 3D dictionary by adding the definition (in your own words) using five words related to 3D printing in this chapter.
2. Print one of the objects (shell, sphericon, or puzzle box) showcased in this chapter and summarize the challenges and the potential issues.
3. Out of the five potential issues (orientation, shirts/brim, rafts, infill, and layer height) take two issues and summarize the problem and solution.
4. In your own words, explain three stress-stain analyses and when you would use them.
5. Summarize the preferable orientations for the Magboot object and why.
6. If you make a shell, sphericon, or puzzle box, describe your process and share the information.

By Nina Bercacio

# Refining and Finishing the Print

## OVERVIEW AND LEARNING OBJECTIVES

**In this chapter:**

- 13.1 – How can I refine and finish my print?
- 13.2 – What are some tools I can use?
- 13.3 – Are there safety and ventilation concerns?
- 13.4 – What are some common practical refinement steps?
- 13.5 – How can I clean my print?
- 13.6 – How can I refine the surface?
- 13.7 – How do I paint my print?
- 13.8 – Additional techniques

## › 13.1 – How can I refine and finish my print?

After your object is completed, it might have the finished look you want or you might need some refinement to achieve the finish you are looking for. In this chapter we showcase a few tips on how to clean up, remove gaps, and paint 3D your prints.

Common refinement steps include:

- **Remove a raft or skirt with pliers** – allows you to control the pull or use leverage and not damage the print.
- **Remove wispy strings with a heat gun** – use high heat and low fan speed. Used very sparingly, this will curl/remove the strings but not have enough time to melt the print – be extra careful near details.
- **Remove irregularities, surface errors, ream holes with a knife or deburring tool** – many of these small irregularities will just pop off; others may need some trimming.
- **Fill the layers** – this includes ABS slurry, wood filler, spray automotive filler primer, automotive filler putty, spot putty, epoxy sculpting material.
- **Sand the surfaces** – best to do by hand so you can tell if the print is getting hot. A mechanical sander can melt the surface too quickly.
- **Paint** – most paints will work.

There are many ways to finish a model; some methods are general, others only work with selected materials, or are more difficult to use, or are time-consuming. But also, ***how you print*** makes a difference. Generally, with a thicker layer height, your object has less fine detail and the layers will be more visible. When

---

**BRIEF REVIEW OF TERMS**

**Shell/Wall** – the outer wall of a model. Shell thickness refers to the number of layers that the outer wall will have between the outside and the infill (think of this as the skin of the object). The higher the setting is for shell thickness, the thicker the outer walls of your object will be.

**Infill** – a measure of how much material will be printed inside the shell of the model. Fill density is usually measured as a percentage of solid. This means that if 100% fill density is selected, the printed object will be solid, with no empty space inside the outer shell. Likewise, if 0% is selected, the printed object will be empty inside. A typical infill is around 20%–40%. The lower percentage saves material and speeds up the print, and higher percentage can help top layers.

you choose a thinner layer height, a higher level of detail is possible and your layers tend to blend better.

For example, if you print at a very large layer height, expect more effort to finish it. Alternately, a very fine layer height often requires much less finishing work. This should be weighed against the print time, as more layers equal more time (often two, four, or six times). The sanding and filler drying time can also add up.

You should factor the model detail level into the equation. When your model has a lot of fine or protruding detail, sanding or other smoothing techniques may physically damage or soften the detail. You may want to print these at your printer's highest layer height to preserve the fine details.

Because of the potential of damage from sanding, you may want to print another shell (i.e., three *shells* versus two) to help make the surface more robust and allow for removed material. Your *infill* settings will also have an effect on the top layers.

> A happy medium is layer heights of .15 mm and two to three coats of filler plus sanding. This seems to keep each stage as short as possible and gives good results. Feel free to play with this and discover your own preferences.

**MAKER'S NOTE**

## › 13.2 – What are some tools I can use?

Before you can start the process, gather the tools and supplies for the finish you desire. Having the right tools for the job can refine your object more effectively (see **Figure 13.1**).

### Common Post-processing Products

- Sandpaper (80, 120, 200, 240, 1000, 1500 grit)
- Handheld electric sander (or sanding block)
- Heat gun
- Plastic spatula
- Mini hobby knife and extra blades
- Needle nose pliers
- Flush cutters
- Chisels
- Water container (for wet sanding)
- Primer/filler spray paint
- Deburring tool
- Sanding block
- Flexible plastic putty knife

FIGURE 13.1 – An assortment of finishing tools. Jonathan Torta.

## Safety Supplies

- Respiratory mask
- Eye protection
- Work gloves

## Common Possessing Products

- Bondo 261 Lightweight Filler Pint Can
- Elmer's Product P9892 Probond Woodfiller
- Apoxie Sculpt modeling compound
- Rust-Oleum Automotive 260510 2-In-1 Filler and Sandable Primer Spray
- Smooth-On XTC-3D High Performance 3D Print Coating
- Krylon K01303A07 Crystal Clear Acrylic Coating Aerosol Spray
- Presto 06006 Kitchen Kettle Multi-Cooker/Steamer

### BRIEF REVIEW OF TERMS

**Bondo** – a brand name filler putty from 3M. It is used for vehicle and household repair, and by artists for sculpting and hobby projects.

**Epoxy** – or polyepoxides, is a resin used for adding a coating to surfaces. Made up of reactive prepolymers and polymers, epoxy reacts and binds (also known as curing) to the material it is coating.

## › 13.3 – Are there safety and ventilation concerns?

Anytime you spray, sand, or use materials with fumes (like *Bondo* or *epoxy*) you must understand the safety issues and have good ventilation and skin and respiratory protection. Read all safety instructions before use. Additionally, power tools, knives, or even the plastic itself can cut or puncture the skin. Take appropriate precautions when handling tools and materials.

## ›13.4 – What are some common practical refinement steps?

When finishing a print, you must know what plastic it was printed in. This can drastically change what finishing methods you can use. However, some common post-processing steps include:

### Basic removal of extra material

**Supplies:** Needle-nose pliers or flush cutters and chisels

- Use tools to remove rafts and support material from the model
- Start large and move to small
- Take care around fine details that may be damaged by rough handling

### Removing wisps (optional)

**Supplies:** Heat gun, plastic spatula

- Scrape and pull off wisps with a plastic spatula
- Fingernails also work well and will not damage the surface
- Use heat gun to melt off wisps

### Initial surface management

**Supplies:** Knife, deburring tool

- Use your tool to pop off, scrape, or cut any extra material that is not removed initially when removing extra material like supports and rafts or wisps.
- Take care not to damage surface.

## Sanding (optional)

**Supplies:** Gloves, sandpaper of different grits

- Sand surface until smooth. See Section 13.6 for instructions.

## Filling (optional)

**Supplies:** Filling material, gloves, spreading device such as brush or plastic putty knife

- Apply filler as directed for material. For spray filler, use the procedure for spray paint.
- Let dry. Some fillers can take 24 to 48 hours to cure.
- Sand off extra material until smooth. See Section 13.6 for instructions.

## Painting (optional)

**Supplies:** Paint, brushes, respirator, gloves

- Spray or brush paint onto object in thin coats, allowing it to dry between.
- Seal paint with clear coating to protect the color layer. See Section 13.7 for painting instructions.

Let's take a closer look at these tasks in the next few sections.

---

**BRIEF REVIEW OF TERMS**

**Curing** – the hardening of plastic or other substances by a secondary chemical process.

---

## › 13.5 – How can I clean my print?

The first task is to remove any extra material from the object. With FDM 3D printing, this is normally removing the brims, rafts, and supports. If printed correctly, these items should be relatively easily removed. It can be challenging in hard-to-reach areas or when surrounded by delicate parts. Take your time. Normally, if you can get a good grip on a part, it will pop off in one hunk. This is where a selection of needle-nose pliers can come in handy for a good combination of grip and fine control. If the material is quite stuck, you can remove it with a cutter.

## Basic removal of extra material

**Supplies:** Needle-nose pliers or flush cutters and chisels

- Manually remove rafts and support material from the model. Use pliers to pull and snap at the base. Flush cutters can be used to cut very close to the model and chisels can be used to help separate.
- Remove large pieces of support first and then approach smaller pieces and fine details.
- Clean the edges and seams of your model.
- If you have a model in many parts, make sure they all fit and align. Do not remove too much material between parts as this may leave a gap that you will need to fill later.

If needed, fine-tune your support/raft settings for a print. These settings in your slicer application will allow you to control how difficult it is to remove or add a gap between support and walls.

**MAKER'S NOTE**

## Wisp removal (optional)

Wisps are created when the hot end moves to a new location. Normally the filament retracts during a move, but sometimes a very thin wisp of filament is drawn, like a spider web. With some filaments, wisps can be almost impossible to stop entirely. Dealing with these quickly is fairly easy, as they can be scraped off. If you use a plastic scraper, it will not mar the surface.

Alternately, a heat gun works very nicely and quickly, as the thin wisps heat much faster than the rest of the object. Use the heat gun very sparingly and keep the model moving. This curls/removes the strings but does not have enough time to melt the print. Be extra careful near small, thin details. PLA can heat up quickly, so use your fingers to feel if it is getting warm. If so, let the model cool before another pass.

**Supplies:** Heat gun, plastic spatula

- Scrape the wisps off with a plastic spatula or your fingernail.
- Apply heat gun. If possible, set to lower heat and low fan speed for more control.
- Move model very quickly in front of the heat gun. Watch for wisps to shrivel. If they do not, move closer and/or move slowly. You are trying to heat the wisps and not the surfaces.

## Initial surface management

Now that the bulk of the material has been removed, you are down to the remainder. This can be slightly more stubborn to remove. If your fingernail or plastic tool does not remove the unwanted material, you may have to use a metal deburring tool to pop off clumps of material or use a knife or cutter to remove them. Brims sometimes peel off cleanly or sometimes need careful trimming.

Be careful not to cut into the surface. If you are going to sand later, trim off most of the remainder but leave some material, as sanding is a better way to remove the worst offending problems cleanly. Take your time and be careful.

**Supplies:** knife, flush cutters and/or deburring tool

- Examine your print and identify bumps, ridges, and generally any leftover material that needs to be removed.
- Use your tool to pop off, scrape, or cut any extra material. Often this material is relatively easy to remove but takes time and attention to find and remove.
- Remove the worst by trimming. Be sure not to cut into the surface.

**MAKER'S NOTE**

Make your life easier – a well-tuned printer and filament create fewer issues to clean. There will always be some leftovers.

## › 13.6 – How can I refine the surface?

The steps in the previous section may be all you need. A raw but manually cleaned-up object can be very nice. At times, you may wish to refine the object even more, as often the layer lines are unwanted, or you may want to paint the print in a color other than the material used.

There are a few ways to smooth or fill the surface of a model. This also has the benefit of preparing the object for painting.

## Sanding

One of the most common ways of smoothing the surface of your print is sanding. Sanding the surface of a model is a fairly straightforward process, but can take some real patience.

For example, PLA has a very low melting temperature; friction can make the surface gummy and soft if it is sanded too aggressively or quickly.

Be prepared to use sandpaper in multiple grit sizes. Using a range of sizes from a 200 (coarse) grit to 1500 or 3000 grit can add many steps and lengthen the process. However, it is the best way to ensure a smooth finish.

Start with a coarse grit of sandpaper. Use a circular motion with a slow, moderate pressure to remove unwanted material. Feel the model often with your fingers to gauge roughness, uneven surfaces, and temperature. The idea is to get rid of material quickly to start the surface, then additional steps refine the surface further.

## Sand with Coarse Grit Sandpaper (Dry or Wet Sanding)

**Supplies:** electric sander (optional) and 80 grit sandpaper

- Begin sanding with coarse sandpaper.
- The goal is to remove any leftover blemishes from the raft or support material and create an even surface that you will later refine. This process removes the greatest amount of material and can take the most time.

  Note that the surface of your print may start looking dull or change color; the shine and color will return as you move to higher sanding grits.
- You can use an electric sander or multitool. Be sure to use a very low setting to avoid overheating and melting the plastic.
- Between each sanding stage, wipe off the model to clear it of any dust and inspect for it a clean surface finish.

## Sand with Medium Grit Sandpaper

**Supplies:** Electric sander (optional) and 120 to 240 grit sandpaper

- Sand with 120 grit sandpaper.
- Sand with 200 grit sandpaper. Each step should refine the surface more.
- Sand with 240 grit sandpaper. If the surface only needs some smoothing, you can start at this step.
- For any large imperfections that you have missed, return to a lower sanding grit to remediate those areas.

## Sand with Fine Grit Sandpaper (Wet Sanding)

**Supplies:** 1000 and 1500 grit sandpaper, plastic bin, and water

When your model's surface is even and starting to become smooth, it is time to wet sand. This process polishes its surface.

- Dunk the model in a tub. Remember to not overfill the tub of water.
- Using 1000 or 1500 grit sandpaper, sand the model until it is completely smooth to the touch and looks shiny.
- Allow the model to dry.
- Inspect it to ensure you have hit all of the areas you needed and it has uniform surface finish.

## Filling the gaps (optional)

Rather than just removing material, you can fill in the layer lines or gaps. This can work nicely, as sanding an entire object down can be time consuming. What the following techniques have in common are brushing on or rubbing in material that fills in the gaps, allowing you to sand and smooth the remainder. While this adds another step and can take some time to dry or cure, you can shorten the manual work and get very smooth results.

Products like Smooth-ON XTC-3D are epoxies that you mix and paint on (see **Figure 13.2**). It flows into the low areas and ridges and cures. Often a single application is enough, with little to no sanding for an extremely smooth exterior with a very nice result. The down side is it can be messy and has a very short open working time, so you need to work quickly in small batches.

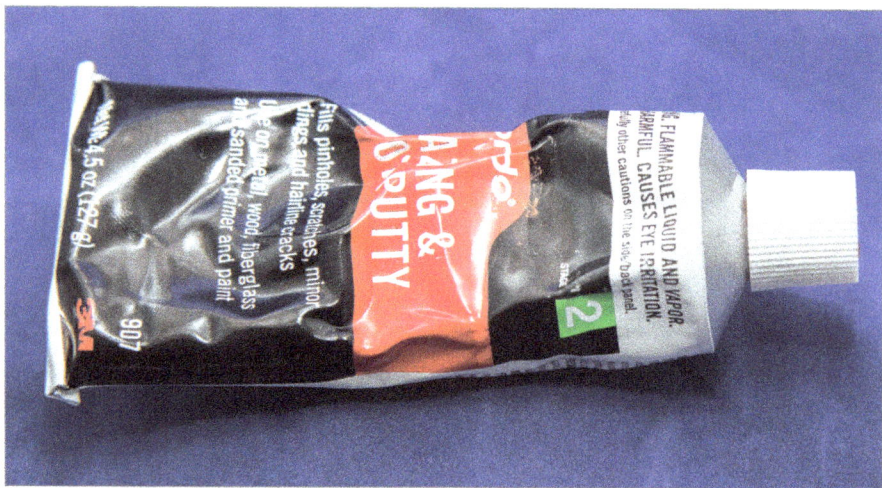

FIGURE 13.2 – An example of Glazing and spot putty used to fill in gaps of a print. Jonathan Torta

You can add various fillers like Bondo and wood filler. These come in different formulations and thicknesses. We like wood filler for the larger gaps and glazing putty for smoothing down surfaces. After they are applied, let the object cure and then sand it (see **Figure 13.3**).

**FIGURE 13.3** – Sanding with filler. Jonathan Torta.

Spray filler works nicely for the surface layer lines. Use in thin layers, let dry, and then wet-sand.

Some common smoothing techniques:

- **Sanding** – basic sanding removal of material
- **Acetone** – only works with ABS; softens and blends surface (acetone vapor bath smoothing is described in Chapter 13.8 with additional techniques)
- **ABS slurry** – make a slurry with ABS and acetone and you can use it as glue or to paint on surface
- **ProBond Woodfiller** – wood filler that works with plastic
- **Bondo Filler** – full Bondo can be useful but is normally way too much for a normal print
- **Bondo 907 Glazing and Spot Putty** – epoxy in a thin, premixed paste – perfect for smoothing the sides of your print
- **Aves Epoxy Sculpt** – an epoxy putty best used for larger gaps
- **Automotive filler primer** – a spray paint filler and primer, easy to use
- **Smooth-On XTC-3D** – clear brush on epoxy, very effective

## Important Tips

When using the filler/primer to fill ridges or gaps, make sure you apply *thin* layers and allow them to dry completely. If the layers are too thick, you may have even more sanding to do and an extended dry/cure time. Ideally, coat the object just enough to see some color from this coat and can see the last layer, then repeat in small amounts after drying. Do not try to cover it fully, as it is much too easy to go from full coverage to over-coverage in one session. It is best to work in steps.

Make sure the object is completely dried or cured. For example, any areas that are tacky to the touch are not completely dry. It can appear dry to the touch, but when sanded can load the sandpaper very quickly. It needs additional curing or drying time.

**MAKER'S NOTE**

Note that temperature and humidity can have a large impact on the drying time.

You can tell that the object is completely dry or cured when you sand and the filler comes off powdery. Another indicator is when it loads a bit on the sandpaper, but comes off easily with a smack or scrape.

Sand the print smooth and check it with your fingers. You can detect a bump only microns thick by touch. You should be able to see the filler/primer in the low areas and the bare plastic in the high. Most often this will be the layer ridges. The print is ready when you can see the layer patterns, but cannot feel them.

From here you can add a thin layer of normal primer or undercoat and finish it with your color. Remember, if there is a blemish, you can re-sand that area (with very fine grit, and very lightly) and repaint. Repeat until you achieve the finish you want.

## › 13.7 – How do I paint my print?

When painting your object, there are a few steps that result in a good, even color and a clean finish. Note that while these instructions are geared somewhat for spray paint, most of the instructions can apply to the brush method.

### Painting basics

The painting process involves four distinct steps:

1. Set up the painting workspace
2. Prime or undercoat (optional)
3. Topcoat
4. Clear-coat

**Supplies:** Tack cloth, fine grit sandpaper, paint

## Painting Workspace and Environmental Factors

If you do not have a well-ventilated room and plan to spray paint, you must do it outside for safety. Unfortunately, this makes you subject to the weather. If it is very humid or cold, you should wait for a better time. The basic rules are that temperatures should be between 50° and 90°F, and relative humidity should be below 85%.

Some paints which are painted by brush, such as acrylic, can be done inside on the kitchen table with a few newspapers.

Make sure the location and environment are appropriate to your paint. Have the appropriate safety precautions in mind, like ventilation, approved respirator, and nitrile gloves while priming and painting.

The painting area should be well-ventilated, clean, and well-lit. For example, wind, dust, or animal hair and flying insects can be an issue. Once it is painted, store the model in a safe, dust-free location as the paint dries.

When priming and painting an object, it is best to first mount it on a painting block or stand using a dowel. This is a useful painting aid. The block or stand can be any object that holds the model securely while allowing the most access to the surface. I have used wooden blocks with dowels, 12-gauge wire, armature wire, and a thin block with nails (think bed of nails) as painting aids. If you have space, you can even hang them. As a best practice, I always use two strings to mitigate all the spinning.

Additionally, you can use parchment paper on a flat surface or aluminum foil that you can shape into a painting stand. Also consider a Lazy Susan or turntable that allows you to spin the model easily, without touching it.

Consider constructing a simple spray booth by hanging plastic sheeting around three sides of the painting area. This will contain the paint to the area around the object. It will also reduce the amount of dust and debris that could adhere to the object while it is drying, or issues from wind if you are outside.

## Primer/Undercoat (optional)

Priming the surface prepares your print for painting. Clean and dust the object to remove any dust or oils. Then use a light coat of primer, remembering to keep it even. This primer paint has more adhesion than standard paint.

Primer is a special type of paint designed to provide a uniform surface that a topcoat paint can easily bond to. Primer comes in both brush-on and spray-on varieties and comes in neutral colors. When choosing a primer for your model, you are better off going with a spray-on variety.

The spray primer covers the surface of your object with an even coat and eliminates the use of brushes, which can leave visible brush marks. Be sure to use a primer and paint that are compatible with plastic.

Applying an undercoat allows you to achieve a rich, deep color by blocking the neutral flat color of the primer. Thus, the undercoat layers are usually either black or white. Use black for darker toned top coat colors and white for lighter toned colors. The undercoat gives the finish extra luster and depth, as it will partially affect the top coat layers.

Apply the undercoat in a thin layer, as noted above. You may need to repeat it a few times. Make sure you have full coverage for the next step (see **Figure 13.4**).

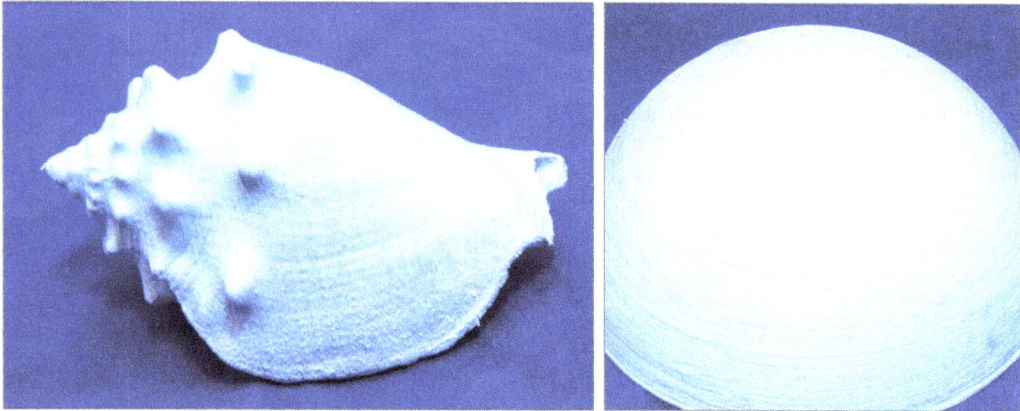

FIGURE 13.4 – The shell print with an undercoat. Jonathan Torta.

Also, there are primer/undercoat combinations that can perform double duty for your color coat. Many paints are formulated to adhere to plastics, so a separate primer is not required. However, it is still recommended. With the primer/undercoat combo you can perform both steps at once.

## Buff and Polish the Primer Coat

An easy way to do this is to use nail-buffing sticks used in manicures, available from any drug store. Use the various buffing and polishing surfaces to buff the primer coat to a glossy shine.

Alternately, use fine grit sandpaper and slow circular motions where possible. The goal is a uniform and smooth surface. Use a tack cloth to gently remove any dust created during the buffing and polishing.

Once you've buffed and polished the primer coat to your satisfaction, you are ready to begin painting.

## Topcoating

The topcoat adds the overall color to your model. This layer can be painted on as you would with the undercoat.

If you used an undercoat:

- You will be adding gradients of color to the undercoat, giving more of a dimensional look and feel
- Perfect coverage is not required as you are adding a layer effect

If you used a topcoat only:

- A topcoat is all you need if you just need color
- Make sure to have full coverage as you do not want any primer showing – remember that it is better to have multiple thin layers rather than fewer heavy layers

Most primers will stick to the main plastics used in 3D printing. In turn, most paints will stick to the primed surfaces. If you prime the printed objects, you should be fine to paint with the paint of your choice.

Additionally, paint washes (paint diluted with water or alcohol) or colored wax rubs can be used to enhance the color layer. It is a good practice to seal them with a clear-coating like Krylon Crystal Clear Acrylic Coating.

See **Figure 13.5** for a finished look at a shell print with a top coating.

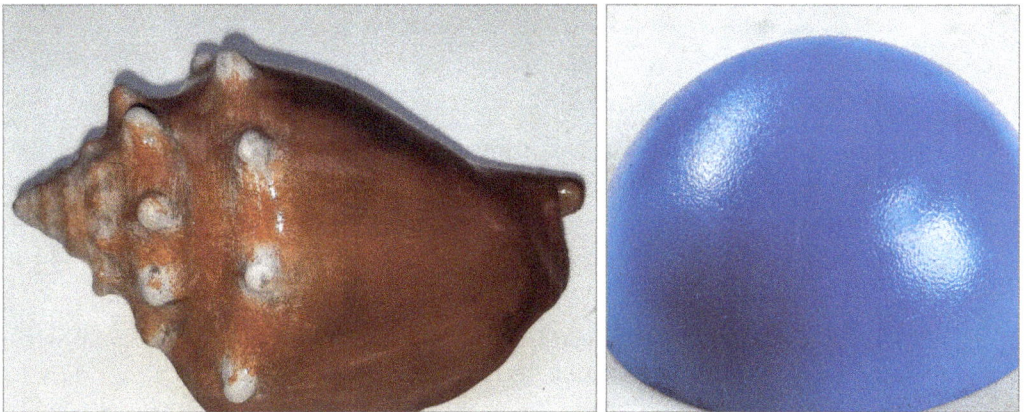

FIGURE 13.5 – The shell print with a top coating (color). Jonathan Torta.

## Clear-coating

Clear-coating is the last step in the painting process. The purpose of clear-coating is adding a final sheen and specular to the model. It also protects the topcoat from damage.

There are a few types of clear-coating available:

- High gloss
- Glossy
- Satin
- Semi-gloss
- Matte

Select the finish you desire for the final appearance. The choice is aesthetic but important for the overall look. Clear-coat is applied in the same way that you have applied every other coating.

You have now finished your painting (see **Figure 13.6**).

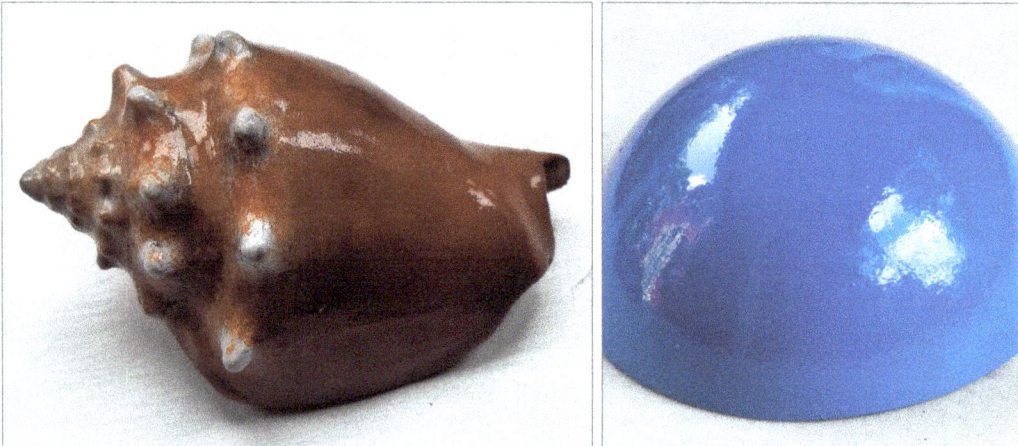

FIGURE 13.6 – The painted shell print with a clear-coat applied. Jonathan Torta.

# Brush painting

You can paint with brushes or airbrushes as you would any other object. However, use a primer because various paints, like acrylic, may not attach to the printed plastic.

# Spray painting

Painting with spray paint can be a little trickier and requires additional steps. Let's take a closer look at some techniques.

## Best Practices for Spray Painting:

### Read the instructions

The instructions on the can are valuable. Take a moment to read them. The spray paint instructions should include a recommended distance to hold the spray can, how long to shake the can before spraying, the proper temperature range, and so on. There should be recommendations on what the paint will ideally adhere to, its solvent or thinners, and the paint's drying time.

### Use a booth and/or drop cloths

As mentioned above, a paint booth, even a crude one, can be a great help. Give some thought to how you make it. Taped-together newspapers are not a robust or reusable solution. A plastic or cloth drop cloth might be easier to set up and clean.

### Use the correct distance

If the instructions do not specify the correct distance, start with the nozzle of the can about 15 to 20 cm (6 to 8 in.) from the surface of the model.

### Do a paint and pattern test

Spray a test shot or two onto a scrap to see the spray pattern that the can produces and the amount of paint laid down in each pass. You can then adjust your technique as needed. Also, this can help you determine whether you need to replace your spray can.

### Sweep back and forth, don't point and blast

To produce an even coat of paint, sweep the can horizontally and vertically past the object as you spray. Don't point and shoot at your model. Use your entire arm to move the can, not just your wrist, and be sure to start the spray before reaching the surface, and release after passing it. For example, if you're moving left to right, begin spraying to the left of the object, onto the object, and then to the right of the object.

### Set the model up high

Set up the model on a platform instead of on the ground. If you set it on the floor, it is harder to see, spray past, and access the lower parts of the model.

### Rotate small objects

Whenever possible, place the object you're painting on a platform that you can rotate as you paint without touching it. If available, set the platform on a turntable like a Lazy Susan.

### Use thin coats

Start with a very thin initial coat that can be built up over time. Do not over-spray your model; it may cause drips, pooling, or uneven coats.

### Let it dry

It is very important to let the paint dry fully. Only add a second coat after the initial coat dries fully.

---

**MAKER'S
NOTE**

---

Some sources say to swirl the primer or paint rather than shaking it. The aim is to not introduce bubbles into the paint. Other sources claim that blasting away for a moment before aiming it at your model is sufficient. Try these and find what works for you.

## Case study: Painting 3D prints

We talk with Nina Buccacio about the painting and coating techniques she used when painting a 3D printed 4-foot dinosaur head.

---

**INTERVIEW**

### JEFF AND NINA BUCCACIO

*Buccacio Sculpture Services LLC*

buccaciosculptureservices.com
www.instagram.com/buccaciosculptureservices/

**ST: How did you become interested in painting 3D prints?**

NB: Even though we focus on sculpture, metal casting, and restoration to public and private art, we welcome many other jobs that require our skills. From the initial design sketch to the finished bronze or paint job to a 4' dinosaur head, Jeff and I work together as a team to execute any given project.

**ST: Did you have to smooth the layers of the print?**

NB: We received the 4' 3D print made of EPS (expanded polystyrene) foam that was already smoothed and was coated with the company's signature coating, then sealed with an acrylic base paint for us to efficiently start in on the paint job as soon as it was dropped off to our location.

**ST: What painting method and paint did you use?**

NB: The reason why we requested the dinosaur head be painted with an acrylic base coat is because I only use the best quality paint on our projects. For this particular 3D print I went with Createx airbrush paint and sealer. Most of the head was painted with

FIGURE 13.7 – The preparation and painting process of a 3D printed dinosaur head. Nina Buccacio.

many different style Iwata airbrushes. There was not much detail to accent with paint on this particular print so I had to actually paint much of the detailed scales in, once the base colors were on (see **Figure 13.7**).

**ST: Were there any challenges in painting the 3D print?**

NB: The print lacked the detail and definition that could have expedited the paint job. The time frame was extended a bit to obtain the desired look from the pictures that were given to us for reference.

Timing is always a factor. With the drying times of the different paint colors applied and inter-coat sealers so colors would not blend but look layered, the whole dinosaur head took about a week from matching color swatches that were provided to us to the final create matte sealer. Then, let it sit overnight to dry before pick up.

**ST: Did you put a finish over the paint?**

NB: We did apply a special high gloss sealer, industry secret, to the eyeballs to give it a glass look (see Figure 13.8).

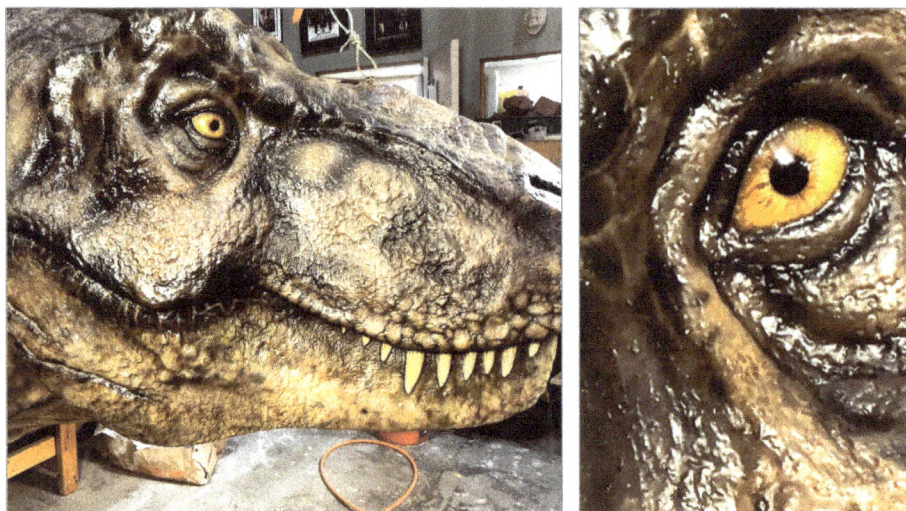

FIGURE 13.8 – Finished painted 3D printed dinosaur head. Nina Buccacio.

## › 13.8 – Additional techniques

Depending on the material you are using and the finish you want to achieve, there are several other specialized techniques that you can use. In this section we list a few examples.

Wood filament contains small wood particles; wood stain can be used to stain it to enhance the look. Use the same technique as you would with wood – you may need to apply more as it may not soak in as much, and definitely seal it (see Figure 13.9).

FIGURE 13.9 – A stained wood PLA frog. Jonathan Torta.

You can also dye most 3D printed materials with advanced liquid dye made for synthetic fibers such as polyester, acrylic, acetate, nylon, and so on. A good example of this is Rit DyeMore, with dozens of colors. Just keep in mind the properties of the plastic you are dyeing. For example, PLA has a rather low softening point, so you may need to keep the temperatures lower while it's in the dye bath.

## Smoothing techniques for ABS

### Acetone Cold Welding

Cold welding refers to the use of acetone to glue ABS parts together. The most common application is to combine multiple parts to make a much larger print, especially when you have a smaller workspace. ABS tends to warp more at larger print sizes. Also, you can take leftover ABS and place it in a jar with some acetone to make a slurry. This can be used to fill gaps and voids.

The easiest way to weld parts is to brush or swab to lightly wet the surface of the model. When the plastic begins to soften, you can join the pieces.

Remember, acetone is not a glue – it softens or even liquefies ABS; use it sparingly. The end product is welded using the original material, so it will be strong.

### Acetone Vapor Bath Smoothing

Vapor smoothing is a variant of welding. It uses a solvent vapor to soften the outer shell of your print with the effect of smoothing it. Acetone can be applied to ABS and most common plastics. PLA is most efficiently soluble in tetrahydrofuran (THF). Polycarbonate is soluble in dichloromethane. PVA is water soluble. HIPS is soluble in Limonene.

Basically, you place your model into a container with the solvent/smoothing agent. The acetone condenses evenly on the part, causing the surface to soften and even liquefy. Surface tension then smooths the plastic. When the model is removed from the chamber, the acetone component evaporates, leaving a very smooth part, free of blemish and visible layer edges (see **Figure 13.10**).

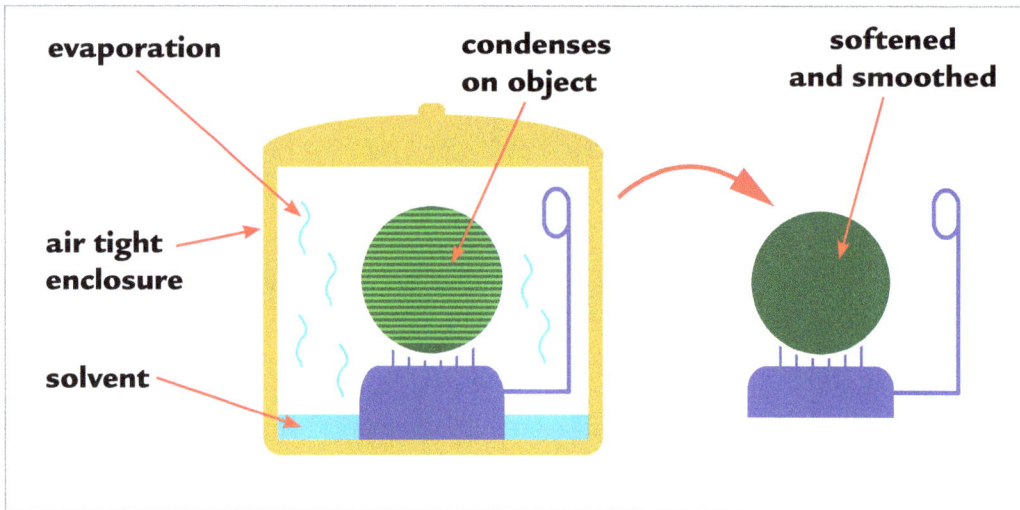

FIGURE 13.10 – Solvent evaporates and condenses on object. This softens and smooths the exterior of the object.

This can be a more complex process than most others, and requires additional safety checks.

## Important Safety Notes

- Vaporizing any chemical solvent is potentially hazardous.
- Many solvents are flammable. You need to know the solvent's boiling point and flame point beforehand.
- Any heating of a solvent *must* be done under controlled conditions.
- Always work with a fire extinguisher handy.
- Solvents are more readily absorbed by the body through respiration and contact.
- Use in a well-ventilated environment.
- Always use a respirator, protective clothing, and gloves.

## Acetone Vapor Bath Process

1. Use a glass or non-reactive metal container for the solvent; for example, a wide-mouthed glass container with a lid or a simple saucepan.

2. Place the model on a platform. I like using a bed of small nails to keep the model from touching other surfaces. Parchment paper or aluminum foil may also be used.

3. Line the solvent container with cotton or paper towels. Pour the acetone on the towels, just enough to moisten the towels. This produces the vapors and keeps the coverage consistent. If you put it on the bottom only, it can take time to fill the container and the bottom of your print will be softer and smoother than the rest.

4. Close the container and wait.

5. Check the print periodically. This is where a glass container or window works nicely.

6. Remove the print (by the platform) when it has softened sufficiently.

**MAKER'S NOTE**

The process will slow down when you remove the print from the vapor bath, but it will not stop immediately. As with cooking, you can remove it just before it is done. The print may be soft, so pick it up by the platform only!

If the vapor bath is too slow, you can heat the container gently to speed the process. Be aware that acetone is quite volatile and will evaporate quickly with even a small amount of heat. Do not use an open flame.

I use a cheap kitchen cooker-steamer: the removable steamer basket works well and the glass top allows me to view the condensing vapor. Just flash to warm and the acetone will quickly become a vapor.

## SUMMARY

Refining and finishing a 3D printed project can be accomplished in a wide variety of ways. You must have the right tools and follow safety and ventilation concerns. First, extra materials must be removed and refined. Next, you must decide if you are going to paint the object and use a primer, top coat, and/or clear coating. There are other special techniques that can be used.

## APPLYING WHAT YOU'VE LEARNED

1. Continue making your own 3D dictionary by adding the definition (in your own words) of five words related to 3D printing in this chapter.
2. Discuss safety and ventilation concerns and why they are important.
3. Discuss two ways you would clean a 3D print.
4. Describe the sanding process.
5. Write a paragraph using the four distinct steps in painting.
6. What are the preliminary steps in spray painting and why are they important?
7. What are the preliminary steps in brush painting and why are they important?
8. What additional techniques can be used in finishing your project?

# PART 3

# Knowledge Base

# CHAPTER 14

By Jonathan Torta.

# Buying and Modifying Your Printer

## OVERVIEW AND LEARNING OBJECTIVES

**In this chapter:**

There is a wide selection to choose from when you are buying or building your own printer. In the first section of this chapter, we outline what to look for when you buy a ready-to-use printer or a printer you can modify or build. In the second section, we give an overview of different modifications you can perform on your 3D printer. **Figure 14.1** shows a commercial printer with laptop showing the 3D model of an object being printed.

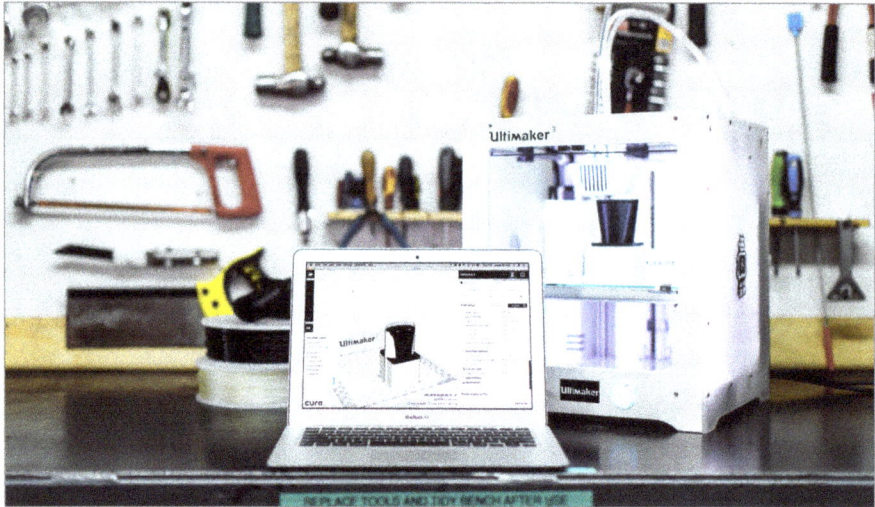

FIGURE 14.1 – A snow machine nozzle model sliced in Ultimaker Cura software and printed on the Ultimaker 3 printer. Svitlana Lozova.

## › 14.1 – Background

Buying a 3D printer can be daunting. Lots of unknowns, coupled with a massive price spectrum, unknown brands, and long lists of features are just the beginning. Also, research information and reviews can also be confusing and contradictory, because the big review houses normally do not handle the newest technology and YouTube reviewers and content can be very biased or uninformed.

Some of this confusion comes from the different tiers and types of printers. As pointed out in Chapter 1, 3D printing is much more diverse than most people realize. A vast array of techniques and machines all fall into the category of additive manufacturing or 3D printing.

When we focus only on FDM printing, the view narrows quite a bit. We now have a single sub-type and there are FDM 3D printers ranging from approximately $50 to $10,000. Though there are many styles of FDM 3D printers, surprisingly, the differences among them are often less an overall factor in buying as most of these variants can give quite similar results, with the main and largest differences being between Cartesian and delta printers.

Cartesian printers are currently the most common, developed, and versatile and have much in common with computer numerical control (CNC) machines. These have a cubic or rectangular build space and are known for their reliability.

Delta printers can be quite fast and almost exclusively use Bowden drives for the filament. They have round build plates and tall form factors. For more information, see Chapter 5.

Both types can make very good prints. I give a slight nod to some direct drive Cartesian printers that may be more versatile, with access to more filament types (very flexible filament can be problematic with some Bowden drives). And a nod to some delta printers for build volume and speed.

## Ways to look at the FDM 3D printer field

The field can be broken down into a few categories: business and industrial, prosumer/small business, consumer, and DIY. These map to price range and overlap greatly. Note that these price ranges are very approximate and are constantly shifting.

Normally, the range below $1,000 dollars is the goal for consumer-level electronics; above that and you start getting to the more dedicated user (hobbyist) level. At a few thousand dollars, you enter the range of professionals and small businesses. Beyond that are the industrial printers. Note that these price ranges overlap quite a bit; price alone is not a perfect indicator except for the extremes.

### Business and industrial

Business and industrial printers tend to be the most pricey and proprietary, ranging from approximately $2,000–$10,000. They are robustly built for continuous and reliable use. They have their own software, support, service plans, and often proprietary filament. They are normally outside the range of most individuals except businesses because of the high upfront cost, along with the higher maintenance cost (for materials and 24/7 service contracts). Note that industrial printers often require a level of expertise from their operators.

### Prosumer/Small Business

Prosumer/small business is a high-end consumer tier of between $1,000–$6000. Normally, they are for small businesses or people who want a smaller-scale, business-level product. These usually have most of the features of the industrial-level products, pared back to make them much cheaper. Prosumer printers are potentially not as robust, large, accurate, and so on. Buyers may still want service agreements and may need reasonable uptime and may not have a dedicated employee or personal expertise to handle repairs. This tier can have its own range of qualities and feature sets, as this can overlap consumer and industrial and the needs of both can differ. As to feature sets, the consumer level is often all about new, cool features, whereas with industrial, it's about reliability. This tier is also very proprietary, generally having its own software and custom hardware.

## Consumer

Consumer is the off-the-shelf, all-in-one, plug-it-in-and-print tier, at a price point of approximately $50–$2,000. It includes a very wide range of prices and capabilities: from tiny "toy" printers, to low-end, low-priced clones, to midrange "brand names," to high-priced feature-showcase models. The commonality here is that these are sold in stores and online and are geared for common people who want to try out 3D printing. Their general focus is on easy printing, hiding the complexity as best they can, and they are meant to be as plug-and-play as possible. They advertise a large feature set and/or the lowest price.

Consumer-level printers can have proprietary or common software and generally are less strict with their filament options (though many are restricted to PLA for ease of support, new user printing, and reduction of parts like the heated bed). They tend have a wide range of warranties and few direct repair services beyond their warranties. A real grab bag of different printers.

This grab bag also extends to the very cheap low-end models that tend to be knock-offs or stripped-down printers made as cheaply as possible. These are no longer easy to use and often fail quickly, requiring an overhaul to even function again. A real buyer-beware range that a non-DIY first time user should stay away from.

## DIY

DIY printers are a wide range of kits, cheap base models (often the very ones avoided above), and open-source fully-capable or experimental printers. The price point varies between approximately $150–$1,000. The focus here is modability and upgrading – the ability to add to and change the base printer. DIY tends to be much rawer, with less need for retail boxes and full instructions, and can include quite a bit of assembly, rework, and tuning.

Ironically, the low-end and clone "consumer" printers that often fail at their job of being a good first printer for a new user are used as base models/frames to be upgraded and even radically changed to a more capable printer by a competent

DIY enthusiast. DIY also includes building kits and making new printers from scratch; this is made easier by the wide array of generic and enthusiast parts on the market, very much like DIY computer building.

DIY is also all about open source, and even higher-end complete printers that are fully open are also quite well received and useful for the DIY enthusiast. DIY is about both the printing and the printer, and like a car enthusiast, the DIYer is constantly tuning, tweaking, and adding to their hobby. Often this also includes a large learning curve and potentially a much greater understanding of the device and process.

When buying, you need to pinpoint what you need, what you want to do, the price range you are capable of, the level of involvement you are willing to invest, and the reliability and uptime required. Answering some or all of these will help narrow the field even more and also give you a good idea what is out there and what you should expect.

If you are seeking an industrial printer – anything less than that is not going to work – simple.

Prosumers value good, reliable output over startup costs. Ease of use and potentially very good service is often a must. A small business needs reliability and good quality prints. The price range is normally much higher.

When seeking a consumer-level printer, you are typically not going to want to open up the guts of the printer to modify/fix something. If there is a problem, replace the unit! You need a good warranty or very easily replaceable parts. The price is a factor, as well as being fast and easy to learn! Print quality needs to be good enough. You should expect just fair prints for the price point and lack of expertise and tuning.

DIY printers satisfy the needs of a build this, mod that, learn the other, enhance this machine attitude. Also, you must have the time and drive to do it all, to learn more about it, and also to progress through difficulties and fix them. You are looking for the best print possible with the tools you have, which could start very cheap and might take some time and effort. The journey is part of the fun!

So what category or tier do you think you fit in? How much can you afford? What plastics and features do you need? How much effect and time can you dedicate?

# Additional suggestions and notable trends

Price indicates a trend of potential quality but does not guarantee it.

For example, cheap guarantees cut corners and low-quality parts but may work just fine – for a time – or be completely unusable. More pricey units could be using better parts but have poor design or packing. That one bad part out of the box could make that quality irrelevant until you can (hopefully) get it fixed. Alternately, the extra money and support might make all the difference.

High-end industrial printers with service contracts need to function well or they lose money and prestige. So they are generally rather solid (though sometimes language barriers can cause confusion). These are not good printers to learn on as they can be much more dangerous and require a fair amount of knowledge to run.

If you are a small business, do not just go out and buy any 3D printer that looks cool and expect someone to become an expert overnight. Many consumer printers may look cool, but they operate with limitations or require a lot of maintenance. It is unfortunately common for a small 3D printer to be gathering dust in a corner after a failed attempt at using it for a grand idea. A good understanding of capabilities and needs as well as expertise is required here.

If you are to pick a DIY gem, you have to understand what is vital and what is not. For example, a strong stiff frame, good stepper motors attached correctly, stepper motors drivers, bearings, rails, and power supplies are things to look for. Quality here saves maintenance and can be the difference between a fair print and a great one. Cheap bearings fail quickly, bent rails can deform prints, loose parts generate poor surfaces or failed prints, low-end drivers blow and make the steppers noisy, poorly tuned drivers and cheap steppers skip, and so on.  This is a steep learning curve but an interesting and fulfilling one.

# Common features to look for

### List of compatible materials

What materials can they handle? Basic printers can print PLA. These normally have no heated bed and are the cheapest as they require the least hardware. Many other materials need higher temperatures, heated beds, hardened nozzles, and even full enclosures. Some flexible materials will not run reliably through some printers. With more flexibility in materials types generally comes more cost. With the wide range of PLA types, many users have no need to use anything else.

## High temperature

In order to use a wider range of materials, a higher temperature range is often required. All-metal hot ends require better tooling quality and are thus more expensive. But they are often a good investment.

## Size of print volume

Size matters; very small printers can only print the smallest of objects or you are forced to cut them up and assemble them later. A current common range is around a 210–270 mm cube/rectangle (often taller). A few printers are now in the 300–500 mm range. These larger printers can do some truly massive prints but can take a week to do so, using many spools of plastic. Also, with size comes scale issues (flexing frame, vibrations, leveling, cost to run the heated bed, etc.).

## Auto-leveling

Manual leveling and tramming can be challenging, and unleveled beds cause many print failures. New automation to calibration processes can not only automatically tram your bed but also map the minute topography and inconstancies of the print bed and correct for it during the print. The down side is not all sensors are reliable. This is a great feature when it works.

## Enclosures

Most early FDM printers printed ABS and required an enclosure to mitigate ABS's tendency to warp. But an enclosure might also be useful to deaden the printer's noise, keep dust and other items out of the build area, and help contain the fumes for a scrubber. Most PLA-only printers do not have enclosures and tend to be cheaper. Cheap DIY enclosures are easy to make.

## Advanced controllers

The first printers had problems keeping up with the G-code during a print. Now many are upgrading to controllers that are small 32-bit computers in their own right. These come with touchscreens, the ability to view and slice objects onboard, Wi-Fi and their own web server to allow remote access and control of the printer (features normally relegated to a laptop or other external controller).

## Advanced stepper drivers

"Microstepping" is where, instead of moving strictly one tooth of the gear or step at a time, the driver can apply just enough current to hold the gear between steps, increasing the accuracy of the output motion. As of today, 1/16th microstepping is fairly standard, and has been for a while, but there are some drivers that can go to 1/32nd, 1/64th, 1/128th, or even 1/256th.

### Color touchscreen

The earliest machines had no control interface whatsoever and any information was seen through a computer interface. That moved to tiny LCD panels and then eventually to large, color touchscreen interfaces, making life a lot easier.

### Networking

Most printers offer USB ports and SD cards for manually delivering or streaming G-code files for printing. Many printers now include wireless and wired networking as default features.

### Filament out detection

A long 3D print will fail when the filament unexpectedly runs out. This is expensive and time-consuming for printer farms, which would have to monitor all of the printers for filament use. For home use, you get the added benefit of being able to use up all the filament on your spool rather than playing it safe and not using the last bit. The printer will detect that the filament is out, stop the print, and save the current location, allowing you to load a new filament and resume the print from where it left off. A real life-saver for some. A rarely run-into problem for others.

### Adhesion systems

Originally painters' tape and Kapton where the mainstays of print surfaces. There are many new surfaces that can cover your build plate and help stick your prints reliably. Some are removeable (or even magnetic) and make handling your prints very easy.

### Heated print surface

Normally a heated bed was only for ABS, but now it is becoming much more widespread as more materials are available. Even PLA can in some cases benefit from low heat. This is a welcome feature if you want to use other materials. Also note that the runtime cost for the printer can be affected greatly by its build plate heater.

### Dual extrusion/multi-filament heads

This adds the ability to 3D print in more than one material (often used for dissolvable support materials). Often this means additional independent extruders. A new trend is using a single hot end that can be fed with more than one filament. These can print with two or more (even five or six) different filaments, and some can even blend them together.

While this is an amazing new set of features, it comes with various downsides: size and weight, as these new hot ends tend to be bulkier and this can affect speed; price, as they are more complicated; and with dual extrusion they need priming with every head change (material is wasted). This is worse with multi-filament heads, as they need to purge the other color from the nozzle before printing the new color. With many color changes this can waste up to two times the filament needed for a single print. Solutions are being worked on and this is a new and exciting feature.

### Power loss recovery

A loss of power can ruin a long 3D print and, like the filament out sensor, this will save the last location until the power is back on and the print can resume. This can save quite a bit of lost time and effort if power is a problem.

## Modifiable printers

If you are a DIY person, you have the most options, whether you are starting a new printer from scratch, buying a kit, *modding* a low-end base model, or tweaking an open-source workhorse.

---

**BRIEF REVIEW OF TERMS**

**Modding** –short for modifications (or modifying), used in reference to the ability or act of enhancing a default object or application for better efficiency and performance. In this case, upgrading the default printer parts that came with a purchased printer.

---

Building a kit is very enjoyable and a nice step between having a completed printer to mod and making one from scratch. You have the parts and instructions, which saves you from having to be an expert on everything before you start. Also, building is a great learning experience and I would suggest anyone try to do it, at least once!

Lastly, there comes a time when DIYers want the ultimate printer and not to resurrect a cheap heap into something useful. They want to experiment and fine-tune a "perfect" machine. This normally comes after you have done all of the above.

This is where a DIY person starts replacing parts – there comes a point where the base unit is not worth using and where many DIYers have issues, spending more money and time upgrading than building from scratch.

**Before you buy, check:**

https://www.youtube.com/channel/UC-skXpTftJJwg5WzLpFLXlA

https://www.youtube.com/channel/UCVc6AHfGw9b2zOE_ZGfmsnw

https://www.youtube.com/channel/UCxQbYGpbdrh-b2ND-AfIybg

**Maybe**

https://www.youtube.com/watch?v=2IE6hSQu4lI

## › 14.2 – What modifications can I make to my printer?

There are all sorts of mods you can make or buy for your 3D printer. Some simple ones that do not affect the prints are: cool decorations, lights, and tool holders. Other modifications can help with printing, like filament guides, cooler fans and shrouds, rigidity supports, dampeners, nozzles, and so on. All the way to full upgrades like new hot ends, beds, or a full enclosure (for ABS).

Thingiverse and other 3D printer object repositories are filled with generic or custom-made parts for all sorts of printers and uses, from the simplest bed clips or bob covers to custom-built cooling fan assemblies made for a specific printer. You can print these and add them to your printer easily.

Other add-ons may need to be purchased, and others require electrical or programming knowledge.

**Web link**

https://www.thingiverse.com/search?q=i3+mod&dwh=495be7593acdefb

Here is a short list of some common mods/upgrades and other add-ons that are useful (if your printer does not have an equivalent or better setup).

- **Remote brains** – Octoprint/Astroprint adds all sorts of useful features (Wi-Fi, remote control and monitoring, viewing, and so on). Some new printers and controller boards have similar features.

- **Auto leveling** – This can take various forms, but "mesh" bed leveling is turning out to be a must- have feature. This takes multiple samples of the entire bed with a sensor on the print head and corrects for any deviations and auto trams!

- **Moisture-sealed filament box** – This keeps your filament in a controlled environment when printing. Most filaments are hydrophilic, but some are much more so than others, and they can become unsubtle during a long print. This is a great way to keep all your filament fresh!

- **Filament guides** – These are simple but can help with potential snags. Also, existing guides may not be adequate or get worn down (see **Figure 14.2**).

- **Vibration dampeners** – these can come in a few forms, but most commonly as added dampener mounts to the stepper motors or rubber feet to the printer. They can make the printer quieter and also reduce vibration that may show up on the skin of the print (ringing). See **Figure 14.3**.

- **Filament spool holders** – Not all spools come in the same sizes. Basic ones may add friction; just one more thing your extruder has to fight. All sorts of options exist – from simple extensions, to ones with bearings, to self-standing ones. Add one to a moisture-sealed filament box and you can use it during the print! See **Figure 14.4**.

FIGURE 14.2 – A new cap for the cold end that holds the ribbon cable and guides the filament. Jonathan Torta.

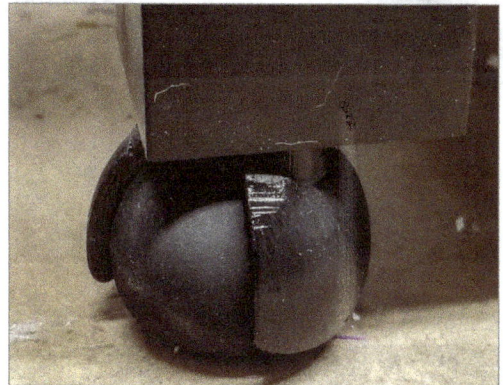

FIGURE 14.3 – Feet for your printer that deaden vibrations (uses squash balls). Jonathan Torta.

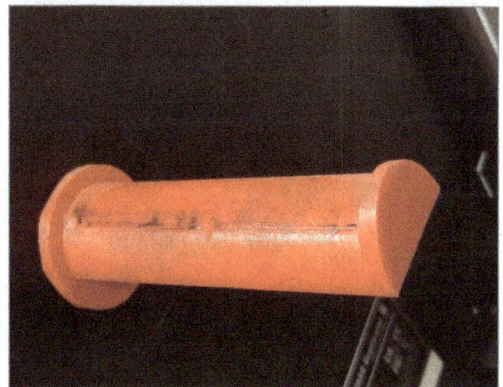

FIGURE 14.4 – A. self-standing filament spool   B. Longer mounted spool holder. Jonathan Torta.

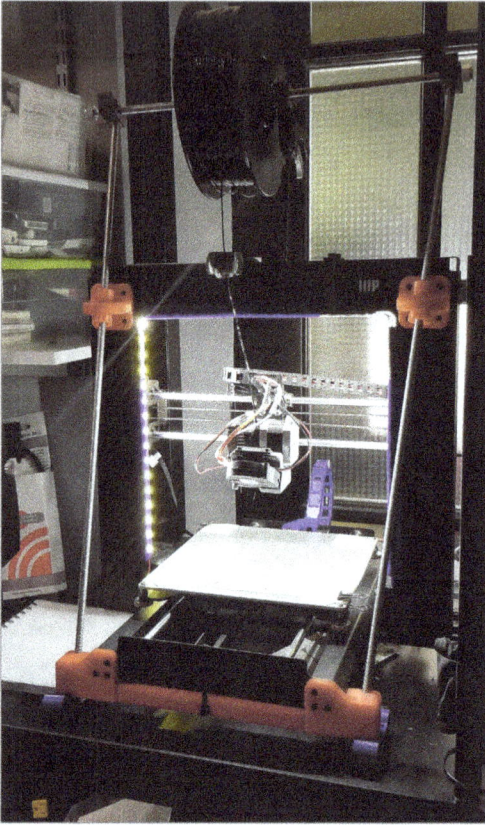

FIGURE 14.5 – Lights, frame stiffeners, and wire management chain. Jonathan Torta.

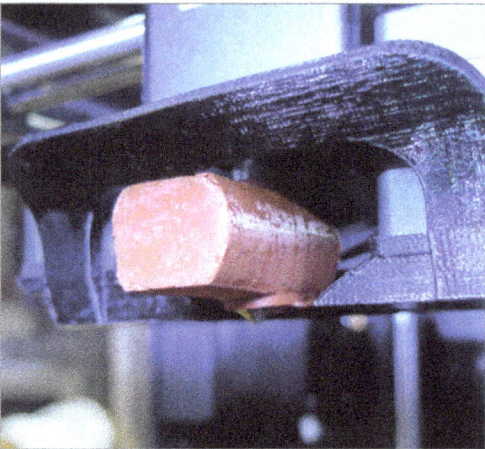

FIGURE 14.6 – DIY boot for the hot end block. Jonathan Torta.

- **Rigidity supports** – It is *very* important for the frame to be rigid. Any movement or vibration will be seen as uneven layers and ringing on the print. Rigidity supports stiffen the frame to remove any extraneous movements or vibrations (see **Figure 14.5**).

- **Hot end boot** – Insulation for your hot end block to keep the hot end warm and protected from filament fan cooling air. These help keep the temperature steady so that the printer does not have to fight itself when the part cooling fan turns on high, inadvertently cooling the block (see **Figure 14.6**).

- **Glass or removable magnetic bed** – A borosilicate (high temperature) glass bed tends to be very flat and also holds heat well. A removable bed is extremely convenient.

- **Additional lighting** – This helps with lighting the work area for you and for your camera. Often a LED strip is used and quite a few printers have these built in if they have an enclosure (see **Figure 14.7**).

- **Cooling fans and ducting** – After it is extruded, PLA printing benefits from good, fast cooling. Unfortunately, not all manufacturers include adequate fans (weak or noisy) or ducting that blows air uniformly around the nozzle (see **Figure 14.8**).

- **Cable chains/relief** – Add these to reduce stress on cable bundles. This will greatly increase the lifetime of these cables, with their constant movement.

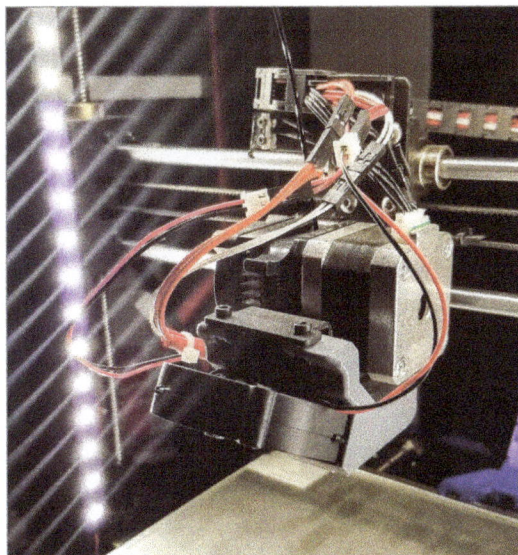

FIGURE 14.7 – Lights. Jonathan Torta.

FIGURE 14.8 – Enhanced part cooling duct. Jonathan Torta.

- **All-steel hot end** – All-steel hot ends can handle higher temperatures. The PTFE/teflon tubes used in many PLA printers will break down at the higher temperatures needed for some filament. An upgrade may be in order to be able to print with these filaments.

- **Hardened nozzle** – If you're planning to print with specialty filaments like carbon fiber, metal, or wood, upgrading your nozzle is important to be able to handle the job. A high quality A2 hardened tool steel nozzle is vital for abrasive filament. And in all, this upgrade will greatly improve the longevity of your nozzle, so it's especially worth investing in for those who use their printer consistently.

IMPORTANT: Only think about modding/upgrading **after** you have calibrated your printer and printed a few times. This ensures the printer works and you have a baseline to compare to. Add one mod at a time and run a test print, noting any changes. If you change too much, or with no baseline, then when you have an issue, you have little idea what went wrong and what to fix. This is a common issue with new DIY modders – they are so enthusiastic to upgrade, they put themselves into a real bind.

MAKER'S
NOTE

## SUMMARY

You must consider a wide range of matters when buying a 3D printer. Are you buying the 3D printer for a large business, industry, a small business, as a consumer, or as a DIY? What do you want it to do? What is your price range? What is your level of involvement? If you want to make modification or upgrade your 3D printer, can you do it with the printer you want to buy? These are just some of the basics questions to ask yourself before spending your money on a 3D printer.

## APPLYING WHAT YOU'VE LEARNED

1. Continue making your own 3D dictionary by adding the definition (in your own words) using five words related to 3D printer in this chapter.

2. What kind of buyer would you be if you were to buy a 3D printer and what would you look for in that kind of printer?

3. Research different 3D printer kits on the present market and write the pros and cons about one of them.

4. What modifications would you want on your 3D printer and why?

By Rawpixel

# Maker Minds

## OVERVIEW AND LEARNING OBJECTIVES

**In this chapter:**

- 15.1 – Collaborating with the maker community
- 15.2 – The maker community

There is a robust 3D printing community and knowledge base available to makers. The collaboration and sharing of the 3D printing community help makers with skills and creativity. **Figure 15.1** shows two makers collaborating during a print.

FIGURE 15.1 – Makers using a 3D printer and laptop. Stokkete.

As we talked about in Chapter 3 and Chapter 4, there are a number of dedicated avenues for collaboration and community groupthink. Online groups, conferences, trade shows, makerspaces, and fabrication labs are just some of the places where makers from all backgrounds and locations can come together to brainstorm, troubleshoot, and learn more about 3D printing.

For example, makerspaces and fabrications are physical community workspaces that share access to manufacturing tools. Makers can go to these workspaces to use equipment they might not have access to. The can also meet with other makers to learn and create projects together.[1] Online forums are virtual communities where makers can collaborate from around the world.

## › 15.1 – Collaborating with the maker community

I contribute to various online groups and follow others closely. I could be anywhere from up-voting good information to offering help and various explanations myself, and even asking questions about items I need more information on. I have found that most groups are very friendly, eager to share experiences and offer any help they can.

Many of these groups also have pointers to documentation, both official and gathered, on a wide range of topics. They share useful 3D models, diagrams, new filament profiles and tests, and instructions for modding, even custom updates or changes to the firmware for the printer (adding a feature or menu item). Some of the most common questions get answered quickly or have their own FAQ or pinned post to avoid repetition. New and interesting issues and problems come up all of the time with common use and exploration.

## Forums and groups
### Community Collaboration and Evolution Story

A new 3D printer I purchased had a common issue with its filament cooler being underpowered and too lopsided. This affected the print quality and limited what it was possible to print well.

While this affected general print quality and I could mitigate it somewhat, printing towers, overhangs, and using bridging was very problematic.

Various users started adding mods or upgrades to supplement the poor cooling. At first, these included upgrading the fan to something slightly stronger, but the available size and types that would fit limited selection greatly and did not help an issue where the air only blew from one side. Then a few users added new air ducts to distribute the air better. Still, more air was required. Later mods added an identical new fan on the opposite side. This helped quite a bit, but was still not enough. We were aggregating a bunch of small incremental fixes that only partially worked.

This is where the community really came together. We all needed a good solution. We put our minds together and broke down the problem, the requirements, and potential issues. When a group of people, including folks from interesting backgrounds and capabilities, got together, various problems and solutions were identified.

For example, there is not much room to mount new hardware – it's limited by collisions with the sides of the printer and the gantry itself. The current mount is on the right side under the heat sink, allowing only a limited space to be used, and it had minimal air duct with a 90-degree bend.

The community started to form solutions to these issues. A duct with multiple air vents, all blowing in from various sides, was deemed the best configuration for the airflow. We debated the number and size of the vents and experimented on them. Meanwhile, with this change, a few people noted that the inline fan, while it can move more air normally when just opened fine, had issues with the backpressure generated by the ducting that constructed the airflow. Replacing it with a blower type fan, even with a lower top CFM (air flow), would create better overall airflow, as it could handle the backpressure created by the more complex ducting and push the air forcefully through.

We created new ducts and fan mounts and tried them out. The new issue became where to put the fan. Blower fans that would fit in the same location as the original fan were rare and tended to very weak because of the necessarily extremely small size. In some other locations the fan was orientated blowing down vertically, but the air had to make a 90-degree turn in the duct. The distances were still quite small, so this was abrupt. It was not ideal solution, but it worked if the blower was strong.

In search of a better solution, I suggested placing the blower on the opposite side. This still was a positioning issue, as the appropriate-size blower fan would not fit horizontally (the ideal position). I extended the mount diagonally. This allowed it to just fit and it had better air flow without the turbulence introduced with a much sharper bend. It also required a few new mount supports.

Many variants later, another member put it all together. We identified a quite powerful, compact blower. We mounted it on the left side at an angle, using all the original mounting holes plus two additional supports with three large air vents, all pointing at the filament nozzle and providing full coverage. We even included a mount for an optional automatic print bed sensor. It was a fusion of the best the community had come up with so far.

The manufacturer's original under-engineered cooling solution was now easily replaced with a user-printable replacement and a new, more appropriate fan. This was not only a replacement fixing a design failure (low, single-direction air flow), but it enhanced the solution, greatly increasing the ability to cool the filament uniformly and allowing for better performance, quality, overhangs, and bridging.

Where previously you had to keep the fan on 100% to get mediocre cooling, the new setup allowed real control over your filament cooling. It allowed you to ramp up the cooling as needed for details, overhangs, and bridging and kept it lower for normal layers, and very low or off for the initial layers. Control, better distribution, and greater range!

This was a community enterprise, with many people looking for a common solution, experimenting, looking at others' work, iterating on good ideas, sharing knowledge, and cooperating. It has not stopped there! There are new variants being made which include alternate mounts for other fans, different mounts for various other automatic print bed sensors, and other refinements.

This is just one example of many. The community is wide and diverse. It is ready to help and make better printers. Some members are specialists in various fields who do this for a hobby; their input is invaluable. Others are avid DIYers with years of experience getting things to work and troubleshooting problems. Even new members can contribute by trying out the mods and upgrades and testing them out and suggesting updates. (see **Figure 15.2**)

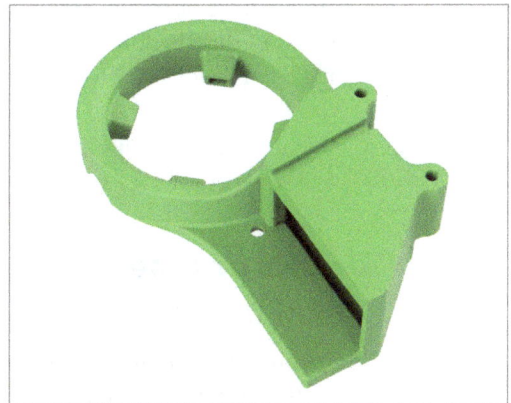

FIGURE 15.2 – Community evolution of a cooling duct. Jonathan Torta.

# › 15.2 – The maker community

There are a lot of online resources a maker can look to for models, help, news, and tips. In this section we will organize some of these resources into categories to make it a little easier to find the information you are looking for.

---

**MAKER'S NOTE** — Because web site links can change over time, the Keyword Search callout includes a list of terms for current web searches.

---

## Brainstorming and collaboration
### One of the Strengths

There are many great forums and social media groups out there with users collaborating and helping each other. A good place to start is your printer's home website. Some have user community forums that talk about issues, fixes, best practices, and advice on that specific printer. They are often moderated by an employee who can help with technical issues. Another good place is social media groups. I frequent groups on Google, Facebook, and Reddit that focus on specific printers, general printing, upgrades and mods to just firmware for the main controller boards. The information ranges from general to very specific, from questions from new users to engineers communing on enhancements.

### General keyword search

- Search for [your printer name] with "club" or "group"

### More general locations

- /r/3dprinting – on Reddit
- 3D printing – on Facebook
- 3D Printing Club – on Facebook
- 3D Printing for Everyone – on Facebook
- 3D Hubs Talk – on Forum
- 3DPrintBoard.com – on Forum
- 3D Printing – on Linkedin
- 3D Printing – on Quora
- 3D Printing – on Google+

## Resources
### Gathering 3D Models to Print

There are many informative online resources to find hundreds of thousands of models you can print with your 3D printer. There are even dedicated search engines that will look though **all** the normal sites with a single search. Most models are freely shared for normal use. Most fall under various Creative Commons licenses. Some sites also sell work/art that users have made, and still others sell their own product.

You can use the lists below as web search keywords:

### Search engines

- Yeggi
- STLFinder
- Yobi3D

### Repositories

- Thingiverse
- Free3D
- Zortrax Library
- GrabCAD
- Dremel Lesson Plans
- 3D Warehouse
- Heroforge
- Sketchfab
- YouMagine

### Marketplaces (free and paid models)

- MyMiniFactory
- STLHive
- Pinshape
- Cults
- Redpah
- XYZprinting 3D Gallery
- NIH 3D Print Exchange
- 3DExport
- 3DKitBash
- 3Dagogo
- Threeding
- Libre3D
- Instructables
- NASA (NASA 3D models)
- British Museum
- Smithsonian Institution and Magazine
- Fab365
- CGTrader

### General keyword search

- STL models
- 3D print models

## Self-guided learning

As we talked a little about in Chapter 3, there are a number of different avenues for self-guided learning.

- Print publications
- Online training
- Conferences and trade shows
- Makerspaces and fabrication labs

## General Online Information

### 3D printing

https://3dprinting.com

http://3dprintingforbeginners.com

http://www.makerspaceforeducation.com/3d-printing-and-design.html

### Makerbot education

https://www.makerbot.com/education

### Thingiverse education project

https://www.thingiverse.com/education

### Adafruit

https://learn.adafruit.com/category/3d-printing

https://learn.adafruit.com/skill-badge-guide-3d-printing/3d-printing-overview

### Maker Shed

https://www.makershed.com/collections/3d-printing-fabrication

https://www.makeuseof.com/tag/beginners-guide-3d-printing

### General keyword search

- 3D printer education
- 3D printer guide
- 3D printing for beginners

## Tips and How-To Videos

Online how-to videos are a great way to increase your knowledge base.

### Maker's Muse

https://www.youtube.com/channel/UCxQbYGpbdrh-b2ND-AfIybg

### Make Anything

https://www.youtube.com/channel/UCVc6AHfGw9b2zOE_ZGfmsnw

### 3D Printing Nerd

https://www.youtube.com/channel/UC_7aK9PpYTqt08ERh1MewlQ

### 3D printing

https://www.youtube.com/channel/UC-skXpTftJJwg5WzLpFLXlA

**3D printing zone**

https://www.youtube.com/channel/UCuwjhSZ1dUKIOO8ZnqYrP1g

**General keyword search**

- How-to 3D print
- How-to [insert your project or printer question]
- adafruit learning printing 3D
- 3D printing within YouTube

## Troubleshooting

**3D printing for beginners**

https://3dprintingforbeginners.com/troubleshoot-3d-printing-problems

**Simplify3D's Print Quality Troubleshooting Guide**

https://www.simplify3d.com/support/printquality-troubleshooting

**Simplify3D's Materials Guide**

https://www.simplify3d.com/support/materials-guide

**Rigid.ink's poster Advanced Overview to Improve Your 3D Print Finish Quality (shown at the right):**

https://rigid.ink/blogs/news/advanced-finish-quality

**Rigid.ink's Ultimate 3D Print Quality Troubleshooting Guide 2018**

https://rigid.ink/pages/ultimate-troubleshooting-guide

**Ninjatek's print quality and troubleshooting guide**

https://ninjatek.com/resources/print-quality-troubleshooting-guide

**MatterHackers' 3D Printer Troubleshooting Guide**

https://www.matterhackers.com/articles/3d-printer-troubleshootingguide

**3DVerkstan's Visual Ultimaker Troubleshooting Guide**

https://support.3dverkstan.se/article/23-a-visual-ultimakertroubleshooting-guide

**All3DP's 3D Printing Troubleshooting guide**

https://all3dp.com/1/common-3d-printingproblems-troubleshooting-3d-printer-issues

**RepRap.org's Print Troubleshooting Pictorial Guide**

http://reprap.org/wiki/Print_Troubleshooting_Pictorial_Guide

**3DVerkstan's Visual Ultimaker Troubleshooting Guide**

https://support.3dverkstan.se/article/23-a-visual-ultimakertroubleshooting-guide

**General keyword search**

- 3D Printer troubleshoot
- [your printer] troubleshoot guide

## SUMMARY

A 3D printing community collaborates and shares its skills and creativity. Many post files so other makers can use, modify, and improve their print. Web site links with key word search call outs have lists of words that are helpful in current websites when you need to do research.

## APPLYING WHAT YOU'VE LEARNED

1. Continue making your own 3D dictionary by adding the definition (in your own words) using five words related to 3D printing in this chapter.
2. Name at least four different ways you can self-learn about 3D printing.
3. Brainstorm with someone about 3D printers and explain what happened.
4. Explain different ways you can collaborate with the maker community about working with 3D printers.
5. What happens when you have a problem with your 3D printer and ask for help solving the problem from the maker community?
6. Explore one of the troubleshooting sites and explain how it might be helpful.
7. Watch an online video on 3D printers and summarize the information.
8. Use one of the key word searches in this book and write about what happened.

## REFERENCES

[1] – https://medium.com/@fab9au/the-maker-movement-a550e68a9ad3

By stokk

# APPENDIX A

## Review Questions

1. Which chapter in this book was the most helpful and why?
2. Which ten words in your own 3D dictionary will be most useful and why?
3. Is there anything else you want to learn about 3D printing?
4. Discuss five ways the information in this book has been helpful.
5. If you did a project included in this book, explain what you did and how it turned out.
6. In the future, what would you like 3D printers to be able to do? Why?
7. Explain what interested you the most about the book on 3D printer?
8. Make a search in your area to find out how 3D printers are used. How can you help their use?
9. Explain why safety and ventilation should be your top priority when using a 3D printer.

## In the Extras and on the DVD

The DVD and downloadable files included with this book contains a number of additional videos, images, a quick reference print checklist, a useful link guide, and practice files. We also included a few additional learning challenges and exercises for further study. The print checklist is useful to post near your printer for a quick reference guide.

IN EXTRAS

For these lessons we will be using the alphabet. We will assume that the letters are extruded into 3D objects (see **Figure A.1**). Also assume all diagonals are 45 degrees or less. And finally to make it harder, cannot be printed on the front or back face of the letter (see **Figure A.2**) which would be preferable in many cases.

FIGURE A.1 – Extruded letters.

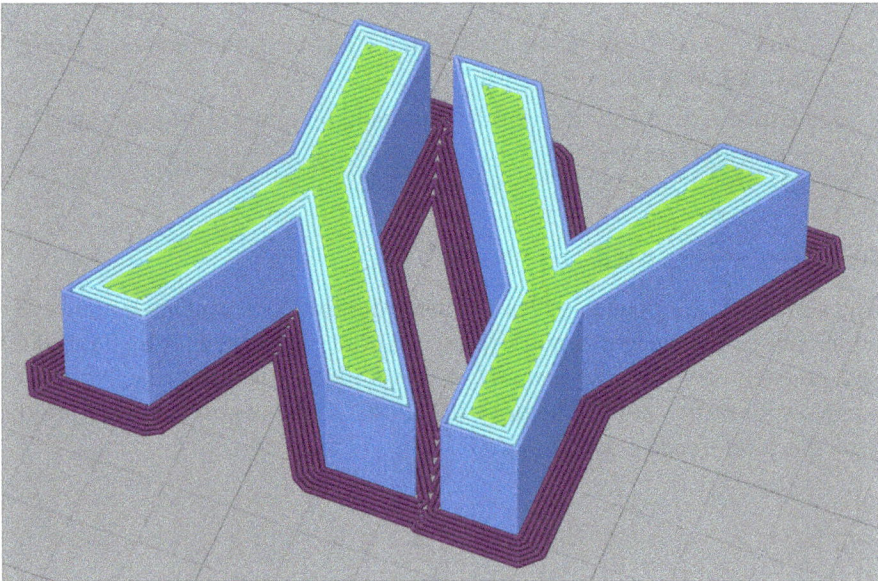

FIGURE A.2 – No laying on front or back.

## Lesson 1

# a B C D E F f G H h K M
# n O P R r S t W X Z

- Select a few letters from above.

- Describe if support is needed?

- If support is needed, why?

- If support is needed, where would it be located?

- Can this letter utilize spanning instead of supports?

- Does your letter have curves and what does that mean for the supports?

**A starter:**

"Y" may not need supports, "T" requires supports, and "A" it may be optional (see **Figure A.3**). Then explain the reasoning.

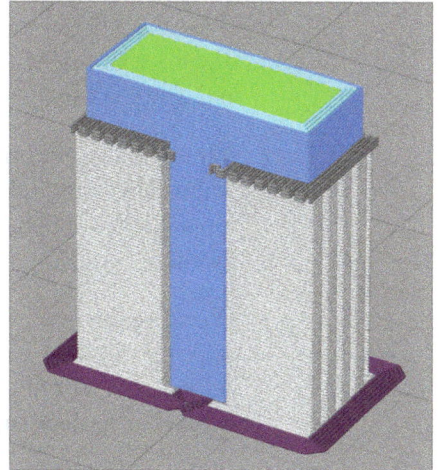

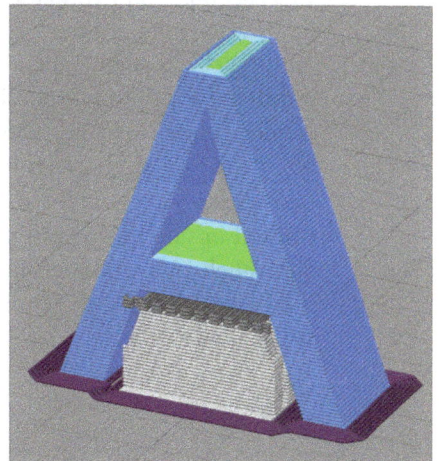

FIGURE A.3 – "Y" standing no support, "T" standing support, "A" standing support optional.

## Lesson 2

For lesson 2, use the letters you picked and described from lesson 1 and continue with the following:

- Is there a better orientation to print the letter that would result in a better/easier print? (remember only rotate the letters right/left not front back for this exercise)

- If there is a better orientation, explain what orientation that is and why you think its better.

- If there is a better orientation, that would result in Less supports? No supports? Spanning?

- Do any of the orientations you came up with above might be helped a brim or raft for potential bed surface support?

For example, see **Figure A.4**

A upright "Y" might need a brim or a raft and if printed on its side would require support, upside down would work with no support and twice the bed surface area. The "T" could be printed upside down with no support and much more bed surface area so this would be the best orientation. What about "A"?

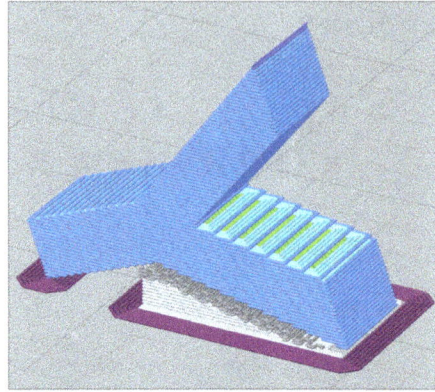

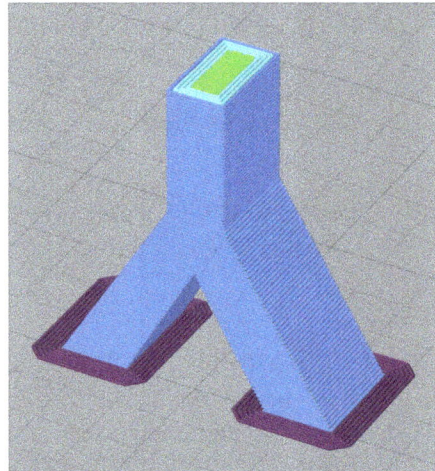

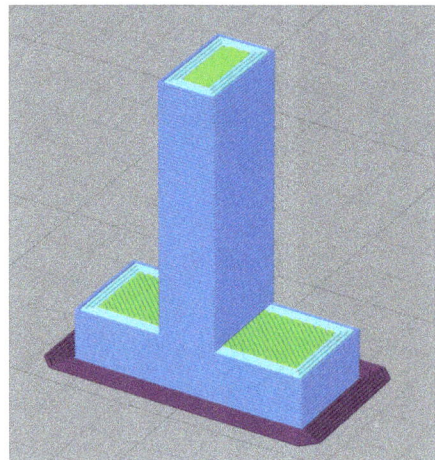

FIGURE A.4 – "y" on side, "Y" upsides down, "T" upsides down.

# Lesson 3

## Questions

1. If I wanted a stronger printed object with least use of additional plastic what slicer permeator would I change and how?

2. If a finished 3D print had Layers are misaligned and shift relative to one another what might be the problem?

3. What additional things you can you do to aid a print stick to the print bed?

4. If a finished 3D print had corners that Curl upward, soft and deformed details what might be the problem? What might be a fix?

5. The layer height increases resolution of what Axis?

6. If a finished 3D print Layers are separate and split apart what is going on?

7. What is a common infill pattern?

8. What is the most common shared 3D printer File format?

9. What are two common FDM Filament types

10. What type of motor moves the print head?

Answers on next page

## Answers

1.  Add additional walls/shells + top and bottom layers; alternative secondary answer: increase infill

2.  Mechanical problems (slipping belts, movement blocked, stepper motors missing steps)

3.  Add brim, raft, print first layer slowly, have a heated bed, use a adhesion aid such as a slurry, glue stick, hair spray or dedicated commercial aid

4.  Overheating, more filament cooling, check environment temps, printing small details too fast

5.  "Z"

6.  Print temperature too low, too much cooling, layer height too large, materials like ABS cooling too quickly

7.  Recliner/grid, triangular, honeycomb/hexagon, zigzag/wavy

8.  STL

9.  PLA, ABS, PETG  - less common but valid TPE, PVA, HIPS, Nylon

10. Stepper motor

By stockasso

# Glossary

**3D Manufacturing Format (3MF)** – data format based in XML and developed and published by the 3MF Consortium; It is used to store model information like shape, color, and materials; additionally, it can store extended information like multiple models and printer settings

**Acetone vapor bath smoothing** – a technique of placing the printed part in a sealed container with acetone in it so the surface softens, and ideally leaving a smooth surface

**Acrylonitrile Butadiene Styrene (ABS)** – a filament used to make durable parts that need to withstand higher temperatures, was used extensively before PLA and can be temperamental to print with

**Additive Manufacturing** – overall term referring to a variety of fabrication processes that uses a manufacturing tool to create a physical object by adding layer upon layer of material; 3D printers are one subset of this type of manufacturing process

**Additive Manufacturing File (AMF)** – XML based data format that can store model information like shape, color, and material; handy for multi-color printers

**Augmented Reality (AR)** – the interactive experience using both computer generated information and information from the real-world environment; often data or imagery projected over a view of the normal world

**Print Bed** – the build area where the 3D print is made; often a flat platform and can be made of a variety of different materials including plastic, metal or glass; additionally, it can be heated to help with plastic adhesion and warping

**Drive Belts** – a belt that transmits motion from a motor, or shaft to a moving part or machine tool; this allows the stepper motors to be placed in better locations and transfer the rotation of the motor to another location/shaft

**Benchmark Slug** – an object used as a sample and can include information like filament type, temperature, and layer height to assist in calibrating the printer

**Software Beta** – a software development phase after alpha; the software is major feature complete but still under development and testing

**Binder Jetting** – similar to an inkjet printer in how it uses nozzles to drop material onto the print area but the binder material is dropped onto a layer of powdered material (metal, glass, and ceramic) in layers

**Bondo** – brand name filler putty trademarked by 3M, used for automotive, marine and household repairs; sometimes used by sculptors

**Bowden drive** – a separated 3D printer head with a stationary cold end connected by a long tube (called a Bowden tube) to the mobile hot end; this is most often to reduce weight

**Brim** – a skirt that touches the object on the first layer or two –a brim does double duty as a priming outline and some extra surface area to hold an object on the print bed

**Build Area** – overall size of the printable area and how large of object can be printed; this includes the XYZ or width, height, and depth dimensions; some build areas are cubical while others are cylindrical; in FDM printing the bottom of the build area is the print bed

**Building on Demand (BOD)** – the first 3D printed permanent European house in Copenhagen, Denmark created by 3D Printhuset

**CAD (Computer Aided Design)** – software used by architects, engineers, drafters, artists, and others to create precision drawings or technical illustrations; CAD software can be used to create three-dimensional (3D) models

**Cal Cat** (basic test block) – a test block that allows you to calibrate your printer settings – this only exists in the chapter it is explained

**Caliper** – an instrument for extremely accurate measuring of external or internal dimensions

**Cartesian 3D printers** – based on the Cartesian coordinates system in mathematics using three axes: X, Y, and Z to move and position the print head

**Chefjet Pro** – a culinary printer from China; includes a cookbook with many different foods to make

**Cold End** – the drive part of the extruder where the filament is pulled from the spool then fed through to the print head and hot end (two common configurations are direct and Bowden drives)

**Cold welding** – a chemical welding process that uses acetone to glue ABS parts together without the use of heat

**Computer Generated imagery (CGI)** – the application of computer graphics to create or contribute to images in art, printed media, video games, films, television programs, shorts, commercials, videos, and simulators; often used to describe computer generated special effects, 3D models and imagery in movies

**Computer Numerical Control (CNC)** – also called numerical control (NC); CNC is the computer-automated control of machining tools; CNC often denotes milling machines (subtractive manufacture)

**Consolidation Processes** – manufacturing method that uses the fusion of smaller individual parts or objects to bond and create a new larger object

**Cooling Fan** – multiple cooling fans located within a 3D printer; these fans help cool areas including the extruder motor, print head, and newly extruded filament for some types of materials

**Cosplay** – act of dressing up as a character from entertainment media such as video games, movies, comics - this can also include supporting props and seasonal events

**Cure** – hardening of polymer material by a secondary chemical process

**Delta 3D FDM printers** – use a round printing bed combined with an extruder that is attached at three triangular points; each of the three points then moves up and down on tracks, thereby determining the position and direction of the print bed; deltas move the print head exclusively

**Digital Light Processing (DLP)** – variant of the SL process that uses mirrors to reflect the UV light to cure large areas of resin

**Direct Metal Deposition (DED)** – uses thermal energy to melt and fuse material (metal powder and wire filament) within the heated or vacuumed print area

**Direct Metal Laser Melting (DMLM)** – uses a laser to selectively fully melt powder metal material into liquid pools

**Direct Metal Laser Sintering (DMLS)** – uses a laser to selectively sinter/partially melt metal powder layer-upon-later

**Do It Yourself (DIY)** – describes enthusiasts that builds, repairs or modifies items without a professional craftsman

**Draft Shield** – a wall around the object being printed to cocoon the object and keep it isolated from the environment; it is best used with materials like ABS that can warp or split if cool air causes uneven cooling and the printer does not have a full or heated enclosure

**Electron Beam Additive Melting (EBAM)** – uses an electron beam and a vacuum build area to melt metal power or wire filament

**Electron Beam Melting (EBM)** – uses a focused electron beam in a vacuum print area to melt powder metals such as titanium, stainless steel, and copper

**Elongation at Break** – the percentage that the material is stretched at the point where it fails

**End Stops** (one for each axis) – a mechanical or optical switch that is triggered when a moving part moves into it

**Entrepreneurs** – individuals who organizes and manages any enterprise, especially a business, usually with considerable initiative and risk

**Epoxy or polyepoxides** – is a resin used for adding a coating to surfaces – made up of reactive prepolymers and polymers, epoxy react and bind (also known as curing) to the material it is coating

**Extruder or print head** – motorized device that has two assemblies, the cold end and the hot end; the cold end pulls filament and feeds it to the hot end that in turns heats the filament before the material exits through the nozzle into the build area

**Extrusion Multiplier** – also called flow rate; a setting that can be important to fine tune the flow rate of filament out of the nozzle; normally in a percentage of max flow rate

**Filament** – raw thermoplastic material used in certain types of 3D printers like FDM

**Filament spool** – a spool of filament material, this format makes it easy to mount for feed into the FDM extruder

**Foodini 3D food printer** – made by Barcelona-based Natural Machine to create new recipes creatively, tastier, faster, and healthier

**Free hand pen** – 3D printer which is essentially just a hand-held print head; you manually move it in space and build plastic up as needed

**Fused Deposition Modeling (FDM)** – process of depositing continuous heating material in layers to create an object; the most common desktop DIY printer; because this term is trademarked by Stratasys Inc., the term fused filament fabrication (FFF) was created and could be used in place of FDM

**G-code** – computer language that communicates the slicing information to the printer; this includes speed, location and path directions

**Geometric Stiffness** – measurement that depends on the shape of the object and not just the properties of the material

**Harness** – measure of the resistance that the surface of the material imposes against penetration by a harder body

**Heat break** – the interface where hot meets cold; this is made to minimize the heat transfer to the cold end

**Heat Sink/Cold End Fan** – a passive heat exchanger that transfers the heat generated in an area to a fluid medium in this case air where it is dissipated away

**Heater block** – joins the nozzle to the heat break, holds the heater cartridge and thermistor and is usually made from aluminum but can be copper plated

**Heater cartridge** – A heating element that heats up the heater block and in turn the filament in the hot end

**Hero Arms** – made by Open Bionics that creates advanced bionic prosthetic arms

**Hobbed Gear** – is half of the filament drive in the cold end; this has teeth and grips into the filament, creating traction

**Hobbyists** – a person who pursues a particular activity done regularly in one's leisure time for pleasure; in this case individuals or groups that use 3D printers for home use, to learn new skills, and for fun

**Hot End** – part of the extruder where the filament is heated and melted at a desired temperature; includes the Heater cartridge and a nozzle

**Hygroscopicity** – the ability to absorb moisture easily

**Hygroscopic** – absorbs moisture from the air

**Idler gear** – a spring-loaded wheel that provides constant pressure against the hobbed gear to pinch the filament between

**Infill %** – the density of the internal space inside the outer shell of an object; this is often measured in percentages (%) instead of millimeters (mm) like the layer height

**Infills** – various internal structures added primarily for support for top layers and also for strength in a 3D print

**Laser Engineered Net Shape (LENS)** – uses lasers to melt selective area of powdered material that is located in the print area; the layer then solidifies before the next layer is added

**Laser Wire Direct Closeout (LWDC)** technique – a wire-based method of printing

**Layer Height** – the vertical resolution of your print and each individual layer that your printer must lay down - this setting specifies the height of each filament layer in your print; it is often measured in milometers or microns; example .2mm or 200µm

**Leveling** (tramming) – the act of making the Print Bed perpendicular to the print head; it is an extremely important process in 3D printing; that if not preformed can printing very difficult to damaging the printer

**Linear Rod or Rails** – a linear-motion bearing or linear slide is a bearing designed to provide free motion in one direction; there are many different types of linear motion bearings; motorized linear slides such as machine slides, XY tables, and roller tables; in 3D printing these are used to move and guide the print head (or bed) around reliably during printing

**Linus Tech Tips** – covers a wide range of topics and has an extensive coverage of various technologies and digital devices

**Local Controller** (control box or microcontroller) – a small computer that runs and controls the functions of the device; 3D printers use one to run the stepper motors, monitor the sensors, power/control the fans and heat, etc. while also making an interface for external control

**Maker** – subculture representing a technology-based extension of DIY culture that intersects with hacker culture and revels in the creation of new devices as well as tinkering with existing ones; typical interests include engineering-oriented pursuits such as electronics, robotics, 3D printing, and the use of Computer Numeric

Control tools, as well as more traditional activities such as metalworking, woodworking, and, mainly, its predecessor, the traditional arts and crafts

**Maker Coin** – an alternate benchmark slug that also is a creative personalized token that can be shared

**Maker community** – community of makers (see above) that share and collaborates their knowledge, help, news and tips

**MakerFleet** – online factory that allows users from around the world to directly access manufacturing equipment in our factory

**Manifold** – is when 3D model geometry is "water-tight"; has a contiguous and defined exterior and interior

**Material Extrusion** – a process used to create objects of a fixed cross-sectional profile; a material is pushed through a die of the desired cross-section; in this context thermoplastic filament is heated and extruded through a nozzle onto the print area in layers

**Material Jetting** – similar to an inkjet printer but uses Ultraviolet (UV) light to cure/harden the material before the next layer is delivered plus it uses supports during the process with a second set of nozzles

**Maximum Stress** (Max Stress, Ultimate Stress, Ultimate Tensile Strength, UTS, or Stress at Break) – measure of maximum stress at the breaking point of the part

**Mesh** – a 3D mesh is the structural build of a 3D model consisting of polygons; 3D meshes use reference points in X, Y and Z axes to define shapes with height, width and depth

**Microstepping** – instead of moving strictly one tooth of the gear or step at a time with normal operation of a stepper motor, the driver can apply just enough current to hold the gear between steps, increasing the accuracy of the output motion; as of today, 1/16th microstepping is fairly standard, and has been for a while, but there are some drivers that can go to 1/32nd, 1/64th, 1/128th, or even 1/256th

**Modding** – is short for modifications (or modifying) and is used in reference to the ability or act of enhancing a default object or application for better efficiency and performance; in this case, upgrading, adding or changing the default printer parts that came with a bought printer

**Model** or Modeling – the act of constructing a digital 3D design created by software, 3D scanner or digital camera

**MyMiniFactory:** – an online community of makers and designers, offering free 3D files for download

**Nanoparticle Jetting (NPJ)** – uses liquid infused with metal particles; this liquid is deposited onto the print bed in a heated build area; the heat evaporates the liquid, leaving a layer of metal

**NASA (National Aeronautics and Space Administration)** – United States Federal Government independent agency responsible for aeronautics and aerospace research plus civilian space programs

**Non-manifold** – when 3D model geometry is not contiguous like unwanted holes or has errors like intersecting edges

**Nozzle** – small metal die with a determined size hole where molten plastic is extruded into the build area; different size nozzles are used depending on the need

**Object file format (OBJ)** – a format used to store object code data; an OBJ can contain additional information like texture vertexes, vertex normals and UV mapping texture coordinates, is made for 3D printing but common and freely useable, and is most often a common export format between programs

**Ooze Shield** – a single shell of extra material around the outside of your part, intended to catch excess material that may drip from the non-printing nozzle

**Parametric** – A method using mathematical equations to represent the points of an object that can also be used to create 3D models

**Polar 3D printers** – an alternate 3D printer coordinates system, as the head positioning is not determined by the x, y, and z coordinates, but by an angle and length; this means that the bed rotates (R) and the print head moves in and out and up and down (X and Z)

**Polyethylene Terephthalate Glycol (PETG)** – Thermoplastic filament FDA approved for food containers and tools used for food consumption, barely warps, and has no odors or fumes when printed

**Polylactic Acid (PLA)** – type of common FDM Thermoplastic filament; PLA filament is odorless, low-warp, eco-friendlier, less energy to process

**Polymers** – large synthetic or natural molecules having a broad range of properties

**Polyvinyl alcohol (PVA)** – filament is non-harmful, non-toxic, environmentally friendly, and can easily be dissolved in water

**Porosity** – the measurement of empty spaces in the total volume of a material

**Powder Bed Fusion (PBF)** – uses the melting of powder material (can be metal or plastic) to fuse particles together to make an object

**Prime Pillar** – useful for extruders that use multiple filaments through one nozzle; the Prime Pillar is the first thing printed when changing nozzles, to help ensure that the nozzle is primed with the new filament, has purged the previous filament and ready to print before it continues with the print

**Print Volume** – the largest size or dimensions of an object that a selected printer can print; can be a rectangular or cylindrical volume

**Prosumer** – technologically savvy person who buys equipment of high quality for non-professional use

**R & D iteration pipeline** – hypothesis development, market research/focus group, concept development, prototype development/optimization, and product testing

**Raft** – horizontal grid first printed on the print bed that acts as a platform to help adhere the primary print object to the bed and aids with uneven print beds

**Rapid Prototyping** – method that turns a digital design into a physical object using a quicker process, such as 3D printing, rather than traditional method

**Robotic arm-based 3D printers** – An alternate method of moving the print head that can allow more flexibility, mobility, and size and thus can be scaled and positioned more accurately than the other printers, making remote or large works possible but on average cannot move as quickly or as accurately

**Sculptris** – a go-to beginners' digital sculpting program that allows a user to model 3D objects like clay

**Secure Digital (SD) Card** – removable data card for mobile devices that come in a range of sizes

**Selective Heat Sintering (SHS)** – uses thermoplastic powders selectively melted by a heated print head

**Selective Laser Melting (SLM)** – uses a laser to selectively fully melt and fuse powder layers in the print area to create an object

**Selective Laser Sintering (SLS)** – uses a laser to sinter/partially melt a variety of materials including thermoplastic, glass, or ceramic powder

**Self-guided learning** – physical and online resources that are tailored to your learning style, speed and location; it can be collaborating with other makers to learn tips are help problem solve potential issues

**Sheet Lamination** – uses very thin layers of material bonded to one another by alternating layers of material and adhesive

**Shell/Wall** – refer to the number of times the outer walls are traced by the 3D printer before starting to print the infill sections of your design; the side "skin" of the object; the thickness is dictated by a multiple of your nozzle diameter and applies to the sidewalls only as the top and bottom surfaces have a different thickness controls

**Sinter** – A process of using heat or pressure to form or compact melted material without liquefying it

**Skirt** – outline of the bottom layer of your object that is printed first and useful to ensure the plastic is primed and flowing correctly before starting your main object

**Slice** or Slicing – cutting an object into thin stratified layers; in this context dividing the 3D model into many cross sections each a printable layer laid down on the previous

**Small Lot Production** – smaller print runs for testing, initial sales or to raise capital

**Stair-Stepping** – a sampling of a continuous surface into detreat steps; a side effect of slicing and the resolution sampling/larger layer height; the sliced surfaces tend to look like topographic maps; this effect is exaggerated with low horizontal shallow/acute angles and curves

**STEAM** – study of a group of fields that include science, technology, engineering, the arts and mathematics

**STEM** – study of a group of fields that include science, technology, engineering, and mathematics

**Stepper Drivers** – electronics responsible for running the stepper motors and controlling their position and the amount of electrical current fed to the motors; many motherboards have the stepper drivers built in or in separate modules

**Stepper motors** – special brushless DC motors that achieve a high level of precision in small movements and can rotate in increments, giving them precise control over their position and speed

**Stereolithography (SL)** – known as Stereolithography Apparatus (SLA) a registered Trademark of 3D Systems, SLA uses a print area vat filled with a liquid photopolymer resin and an ultraviolet laser and mirrors to cure the resin

**Stereolithography (STL)** – a simplistic file format that holds 3D modeling and slicing information readable by a 3D printer; the format comes in two versions (Binary and ASCII) and contains only the mesh and normal file information – the coordinates have no units but typically millimeters or inches

**Subtractive Manufacturing** – manufacturing fabrication process that cuts or drills away from a solid material block to create an object; industrial CNC (Computer Numerical Control) machining is a subset of this process

**Supports** – printed structures added to the print that hold up 3D objects or sections that don't have enough base material (surface area) touching the build plate during printing or when the overhang is greater than 45 degrees

**Thermistor** – an electrical resistor whose resistance is greatly reduced by heating, used for measurement and control

**Thermocouple** – a thermoelectric device for measuring temperature, consisting of two wires of different metals connected at two points, a voltage being developed between the two junctions in proportion to the temperature difference

**Thermoplastic** – type of plastic that can repeatedly become pliable when heated at a specific temperature and converts back to hard and solid when cooled

**Thermoplastic Filaments** – thermoplastic in filament form rather than common pellet form used most often with fused deposition modeling 3D printers

**Thingiverse** – a website for downloading files and builds them with your laser cutter, 3D printer, or CNC; the website with lots of different resources and downloadable 3D content - invaluable to the beginner

**Threaded Rods** or Leadscrews – used with the printer's z-axis, rotates back and forth forcing a nut to move up or down which in turn moves the print bed or extruder

**Three-dimensional (3D) Printer** – manufacturing tool that creates tangible objects from a 3D model design using an additive manufacturing method by adding layers upon layers of material to create a 3D object

**TinkerCAD** – a simple 3D modelling program that serves as a great introduction for those starting up

**Toughness** – amount of energy that a material can absorb before it fractures

**Tramming** – adjusting the print bed perpendicular to the print head

**Universal Serial Bus (USB) Cable** – the term USB stands for "Universal Serial Bus"; USB cable assemblies are some of the most popular cable types available, used mostly to connect computers to peripheral devices such as cameras, camcorders, printers, scanners, and more

**Vapor Smoothing** – solvent vapor to soften the outer shell of an object with the effect of smoothing it

**Vase Mode** – a slicer-dependent setting that prints one continuous filament, slowing spiraling upwards; limited to basic shapes with no infill and one shell/wall

**Vat Photopolymerization** – uses liquid photopolymer resins in a variety of deployment methods that use light (not heat) to fuse/cure the resins in layers to form the object

**Virtual Reality (VR)** – a simulated environment created using an interactive computer-generated program

**Viscosity** – friction between the molecules in a fluid

**Wi-Fi** – Wi-Fi is the name of a popular wireless networking technology that uses radio waves to provide wireless high-speed Internet and network connections

**Worbla** – brand of thermoplastics designed for sculpting, often found in sheets

**X-axis** – a Cartesian coordinate, the principal or horizontal axis of a system of coordinates, perpendicular to y-axis

**Y-axis** – a Cartesian coordinate, the secondary or vertical axis of a system of coordinates; perpendicular to the x axis

**Yield Stress** ("Stress at Yield") – a measure of strength at the point where the part experiences a permanent deformation of 0.2% of the original dimension

**Young Modulus** ("Young's Modulus", "rigidity", or "Modulus of Elasticity") – measure of the stiffness of a given material (not a particular shape created from that material) -this is basically a measure of a material's ability to resist permanent deformation

**Z-axis** – a Cartesian coordinate, the third axis, usually represents depth, in a three-dimensional system of coordinates; the z-axis is perpendicular to both the x-axis and y-axis

# INDEX

## E

education centers, 56–65
electrical components, 125–126
electron beam additive melting (EBAM), 108
electron beam melting (EBM), 107
electronics, 100–102
elongation at break, 262
e-NABLE organization, 68, 69
enclosure, 124
enclosure fan, 126
end stops, 123–124
entertainment industry, 44–46
entrepreneurs, 76
environmental factors, 293
epoxy or polyepoxides, 284
extended infill settings, 207
extended support settings, 207
extruder, 17
extrusion multiplier, 246

## F

fabrication labs, 72–73
fabrication technology, 65, 66, 68
FDM. *see* fused deposition modeling
FDM printing, 241, 250
filament, 7
  calibrating, 244–249
    temperature, 244–246
    using calipers, 246–249
  diameter and consistency, 242
  formulation of, 243
  moisture as concern for, 243
  quick guide to picking, 240–244
    compatibility and limitations of printer, 240–241
    quality, 241–244
    size, 240
    type, 241
  recovering moist filament, 243

filament guides, 317
filament loading, 214
filament spool, 17
filament spool holders, 317
filament thermoplastic, formulation of, 243
file formats, differences between, 188–189
fill density, 177–178
filling, 286, 290–292
filters, 126, 209
fine control of cooling fan, 207
fine control of temperature, 207
first layer settings, 207
Flashprint, 167
flexible polyester (FPE), 152
flow rate calibration, 246
Foodini 3D food printer, 37
frame, 124
FreeCAD, 160
free hand pen, 112
fused deposition modeling (FDM), 15, 110
fused deposition modeling (FDM) 3D printer
  common basic filament for, 133–140
  components, 114–126
    bed, 123
    electrical, 125–126
    frame, 124
    mechanical, 117–123
    motor, 124
  selecting the materials for, 131–132
  specialty materials to print with, 147–152
  work of, 111–114
Fusion 360™, 164
"fuzzy" exterior, 208

## G

gap, 210–215
gap instructions, 210–215
G-code, 12, 13, 208, 216

general online information, 328
geometric stiffness, 263
glass or removable magnetic bed, 318
glow-in-the-dark PLA, 144–145
*Greatest Showman, The*, 46
groups and organizations, 63
Gulati, Harnek, interview, 193–195

## H

hands-on reproductions, 29–31
HangTime Hook, 83–87
hardened nozzle, 319
hardware tests, need to run, 210–214
harness, 263
heat break, 121
heater block, 121
heater cartridge, 121
heat sink, 121
Hero Arm, 50
high impact polystyrene (HIPS), 138–139
high resolution, 174
HIPS. *see* high impact polystyrene (HIPS)
hobbed Gear, 117
hobbyists, 76
hot end, 12, 13, 17, 119–123
hot end boot, 318
hot end fan, 121
household items, 97–100
household spray nozzle, 98–100
HP® Sprout Pro, 65
Hull, Charles, 18
Hu, Tony, interview, 77–79
hybrid materials, 140–146
hygroscopic, 137, 140
hygroscopicity, 40

## I

IceSL, 168
Idler gear, 117

www.ingramcontent.com/pod-product-compliance
Lightning Source LLC
Chambersburg PA
CBHW080226270326
41926CB00020B/4157